Understanding
Chemical Principals
A Learning Companion

Understanding
Chemical Principals
A Learning Companion

Peter J. Krieger, Ed.D.
Professor of Natural Sciences
Palm Beach Community College
Lake Worth, Florida

Prentice Hall
Upper Saddle River, NJ 07458

Library of Congress Cataloging-in-Publication Data

Krieger, Peter J., 1942–

    Understanding chemical principles : a learning companion / Peter J. Krieger.
      p. cm.
    Includes index.
    ISBN 0-13-681321-6
    1. Chemistry--Mathematics.   I. Title.
    QD39.5.M3K75   1999
    540--dc21

                            98-18150
                            CIP

Publisher: *Susan Katz*
Acquisitions Editor: *Mark Cohen*
Production Editor: *Cathy O'Connell*
Managing Production Editor: *Patrick Walsh*
Director of Production and Manufacturing: *Bruce Johnson*
Senior Production Manager: *Ilene Sanford*
Editorial Assistant: *Stephanie Camangian*
Marketing Coordinator: *Cindy Frederick*
Cover Design: *Miguel Ortiz*
Creative Director: *Marianne Frasco*
Cover Director: *Jayne Conte*
Printing and Binding: *Banta Company*

© 1999 by Prentice-Hall, Inc.
A Simon & Schuster Company
Upper Saddle River, New Jersey 07458

Printed in the United States of America

10 9 8 7 6 5 4 3 2 1

ISBN 0-13-681321-6

PRENTICE-HALL INTERNATIONAL (UK) LIMITED, *London*
PRENTICE-HALL OF AUSTRALIA PTY, LIMITED, *Sydney*
PRENTICE-HALL CANADA INC., *Toronto*
PRENTICE-HALL HISPANOAMERICANA, S.A., *Mexico*
PRENTICE-HALL OF INDIA PRIVATE LIMITED, *New Delhi*
PRENTICE-HALL OF JAPAN, INC., *Tokoyo*
SIMON & SCHUSTER ASIA PTE, LTD., *Singapore*
EDITORA PRENTICE-HALL DO BRASIL, LTDA., *Rio de Janeiro*

To Micki, my source of encouragement.

CONTENTS

PREFACE

Because chemistry is an applied-math science, it is very difficult to perform well in any chemistry course without a firm understanding of math and the ability to apply that math. Of course, word problems are a real headache. Naturally, Chemistry is loaded with word problems.

Understanding Chemical Principles: A Learning Companion has been written with the primary goal of removing the mystery from the math found in chemistry courses required for allied health programs and preparatory chemistry. The terminology used, examples presented, and explanations provided are chosen with a minimum amount of formality. Simple and clear presentations are provided to encourage the understanding of math, problem solving, and principles.

Understanding Chemical Principles: A Learning Companion includes the tricks of the trade to make life a bit easier for students. Some of these tricks are designed to help the reading and understanding of word problems. Other tricks are presented that are shortcuts used to speed up the process of working math problems, regardless of the science you are attempting to master. In addition, demonstration problem examples are solved using different techniques to expose you to various problem-solving methods and prepare you for optimal understanding of problem solving in general.

There are practice exercises in Questions and Problems at the end of each chapter. Problems are written in differing styles so there is exposure to multiple ways of wording the same type of problem. The solutions and/or answers for all of the problems are presented in Solutions and Answers at the end of the book.

Each of the chapters is presented so a chapter is reasonably independent of the remainder of the book. Each math skill that repeats throughout the book is explained each time the skill appears so you will have plenty of opportunities to master the tools required for success in chemistry.

Understanding Chemical Principles: A Learning Companion is written with a minimum of technical vocabulary. The use of terminology is restricted to those terms that appear repeatedly throughout the study of chemistry courses. Each of the chemical terms is printed in capital, boldface letters with a definition the first time it is used. The Index/Glossary is provided for quick reference to the definitions of terms used in this book and includes some terms that are in general use.

Peter J. Krieger, EdD
1999

ACKNOWLEDGMENTS

I extend my heartfelt thanks to the students who have tested this book and have provided valuable comments.

I am happy to have worked again with David D. Duncan, Senior Instructor of English at Palm Beach Community College. Mr. Duncan did a terrific job making certain that this book is readable and student-friendly.

I especially appreciate the time and effort of the reviewers for their valuable critique of this book.

Nancy Calder, MT (ASCP, CLS (NCA), M.A.Ed.
Medical Technology Program Director
Thomas Jefferson University
College of Health Professions
Philadelphia, Pennsylvania

Wayne B. Counts
Professor and Chair, Department of Chemistry
Georgia Southwestern State University
Americus, Georgia

Joycelyn Hulsebus, Ph.D., MT (ASCP)
Assistant Department Chair, Department of Medical Technology
University of Kansas Medical Center
Kansas City, Kansas

Rod Tracey, Ph.D.
Professor of Chemistry, Division of Science and Mathematics
College of the Desert
Palm Desert, California

HOW TO USE THIS BOOK

GENERAL COMMENTS

Read carefully and do not rush. This book is written to aid you in your understanding of the applied math of chemistry. Accessory information is included to help with that understanding. Read through each section completely and then cover the math by working through the examples. Section by section, read and work through to the end of the chapter.

A wide margin is provided for notes. It is a good idea to make notes about each topic and to cross reference the topics to other sources you are using by noting the page or pages that apply. These page notes will help you save time when flipping back and forth between this book and other books.

Watch for NOTES, HINTS, and SHORTCUTS. These comments are included in the chapters to help you understand the concepts and how to work the problems. NOTES that designed to cover information that may be confusing or include comments that are critical to success in chemistry.

An Index/Glossary is included for quick reference to technical terminology. The terms included in the Index/Glossary are used in this book and are common to most textbooks. There are some terms included that are out of date but still in use. These terms are included so you will be familiar with them. Definitions for the older terms are referred to the modern terminology and definitions within the Index/Glossary for your convenience.

EXAMPLES

Follow the examples. Even though the examples may look a bit imposing, each example is worked leaving out *no* mathematical steps. Since all steps and explanations are included, the examples take up a lot of room. Just take your time and read through them carefully. The goal of the examples is clear understanding of the problem types and the math involved.

Read a problem carefully. Identify what the problem is requesting and what information you are given to arrive at the answer.

PROBLEMS

Work the problems at the end of each chapter no more than twice. There is no need to get frustrated. If you feel you have to refer to the Solutions and Answers to work the problem, let the problem "cool off" for a few hours or a day and try again before resorting to Solutions and Answers.

Problems at the end of each chapter are in groups. The first of each problem type has a complete solution; the remainder of each type have answers derived from an identical or similar solution. If you have trouble arriving at the answer to a problem, follow the solution either in the chapter or as covered by the first of the problem type. *Aim toward an understanding of how the problem actually works.* There are no surprises or trick problems at the end of the chapters.

MEASUREMENT

• The Metric System, a Rationale •
• Prefixes • Conversion Factors • Word Problems & Dimensional Analysis •
• Distance • Area • Volume • Weight • Density • Specific Gravity •
• Temperature Scales • Rounding Off an Answer •
• Questions and Problems •

THE METRIC SYSTEM, A RATIONALE

Communication sometimes requires a method of expressing sizes that are easily understood. Misunderstanding the communication of measurements can lead to problems. Let your mind wander to the problem of the early hunter coming back to the village asking for some help in carrying an animal home for meat. Suppose that hunter can only communicate the size of the animal with hand signs and uses a circling motion over his head. The villagers interpret that sign to represent something the size of a horse. The problem is that the hunter found an elephant!

The same kind of problem exists in terms of modern measurements. There are two commonly used methods of measurement: the English and the metric systems. The English system of measurement is based on some interesting history. For instance, the inch and the foot are based on the length of one of the joints of a finger of a long dead English king. The foot is based on the length of that same deceased English king's foot—the right one. The length of a foot is considered to be 12 times the length of an inch; there is no ease when converting between the two types of measurement. There are 16 ounces to a pound, another inconvenient conversion between measurements. As a matter of fact, there are no easily convertible measurements in the English system, but the English system is an attempt to establish some form of easily communicated system of measurement.

The metric system of measurement was initiated in France in the late 1700s and was made the standard of trade. There was no standard before that time. Can you imagine what a mess it was for England, for instance, to trade with Spain when there was no common way to measure? Since France was a major power in Europe at the time, the system became a standard for Europe and, hence, the remainder of the civilized world.

The metric system is in common use in the sciences and scientific applications such as engineering and medicine.

PREFIXES

The metric system is a decimal (units of 10s) system. In the English system, increasing and decreasing units are in threes, sixteens, five thousand two hundred and eighties, or by some other strange conversion factor. Metric system changes are performed by multiplying to some multiple of 10 (10, 100, or whatever) or dividing by some multiple of 10. Simply move the decimal point. Each of the multiples of 10 has a name that is used as a prefix (placed before the word that represents the basic measurement). An example of this system is that a meter is a basic measure of distance (roughly equivalent to one yard) and that 0.01 meters is a centimeter (roughly equivalent to one-half inch). So 5 meters is equivalent to 500 centimeters. Try to make the conversion from 5 yards to inches and you will see that some thought is necessary; the metric conversions are much easier.

Some of the more commonly used prefixes and their values are as follows:

Prefix	Symbol	Number	Value
micro	(μ)	1/1,000,000	(0.000001)
milli	(m)	1/1,000	(0.001)
centi	(c)	1/100	(0.01)
deci	(d)	1/10	(0.1)
deka	(D)	10	
hekta	(H)	100	
kilo	(k)	1,000	
mega	(M)	1,000,000	

The prefixes are used to explain the numerical value of the type of measurement being performed; for instance, 1 centimeter (1 cm) is 1/100th of a meter, 1 deciliter is 1/10th of a liter, 1 kilogram is 1,000 grams.

You may consider each of these prefixes as if it were a number and multiply that number times the basic unit. Suppose you had to convert from 5.0 cL (centiliters) to L (liters). You could do it this way:

If you saw symbols right next to each other, let us say *QRS*, they can be written as $Q \times R \times S$ and then multiplied as in algebra.

"5.0 cL" is made up of three symbols next to each other and can be written as $5.0 \times c \times L$ and multiplied. Remember that 5.0 and c are actually numbers.

Multiplication of $5.0 \times 0.01 \times L$ becomes $0.05 \times L$ or 0.05 L.

The trick is to keep in mind that the metric prefixes are *all* numbers and may be multiplied.

$$18 \ \mu sec = 18 \times 0.000001 \times sec = 0.000018 \ sec$$

$$0.5 \ km = 0.5 \times 1,000 \times m = 500 \ m$$

The prefixes may be combined to represent numbers that may or may not have common prefixes.

$$8 \ m\mu L = 8 \times 0.001 \times 0.000001 \times L = 0.000000008 \ L$$

$$3 \ HMg = 3 \times 100 \times 1,000,000 \times g = 300,000,000 \ g$$

$$320 \ c\mu g = 320 \times 0.01 \times 0.000001 \times g = 0.0000032 \ g$$

CONVERSION FACTORS

For those of us who did not grow up using the metric system, one of the problems with the metric system is getting a feel for the measurements. There are lots of conversion factors that may be used to convert between the metric system and the English system, but to make life a little easier, there are only a few that are required to do the conversions—one each for distance (also used for area and volume), volume, and weight. One conversion *within* the metric system may be used and will be discussed under the heading "Density." Temperature is the last type of conversion required and will be discussed in the section "Temperature Scales."

The conversion factors for distance, area, and volume—three only—are

$$1 \text{ inch} = 2.54 \text{ centimeters}$$
$$1 \text{ liter} = 1.06 \text{ quarts}$$
$$1 \text{ kilogram} = 2.2 \text{ pounds}$$

Notice that 1 inch, approximately 1 quart, and a little more than 2 pounds (a small bag of sugar) are pretty easy to recognize. Of course, you have to know how to use these conversion factors to make the conversions.

The word **FACTOR** has been used often. A factor is a number that can be used to multiply; that is exactly how these conversion factors are used.

Suppose we were to look at 1 inch = 2.54 centimeters and read that in words. We could say, "There is one inch per two point five four centimeters." That kind of wording is used to read a fraction. The fraction would look like this:

$$\frac{1 \text{ in}}{2.54 \text{ cm}}$$

Another way of wording the same conversion would be from the opposite direction and would read, "2.54 cm per 1 inch," and that would look like this:

$$\frac{2.54 \text{ cm}}{1 \text{ in}}$$

Both of these ways of expressing the conversion factor mean exactly the same thing. Look at the ruler in Figure 1.1, that is in both metric and English measurement. You would have little trouble seeing the information for 2.54 cm/in OR for the reverse (actually, the inverse), 1 in/2.54 cm.

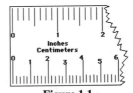

Figure 1.1
Ruler,
English and metric.

The three conversion factors mentioned earlier for distance, area, and volume may be written in two ways each:

$$\frac{1 \text{ in}}{2.54 \text{ cm}} \quad \textbf{OR} \quad \frac{2.54 \text{ cm}}{1 \text{ in}} \qquad \frac{1 \text{ L}}{1.06 \text{ qt}} \quad \textbf{OR} \quad \frac{1.06 \text{ qt}}{1 \text{ L}} \qquad \frac{1 \text{ kg}}{2.2 \text{ lb}} \quad \textbf{OR} \quad \frac{2.2 \text{ lb}}{1 \text{ kg}}$$

WORD PROBLEMS AND DIMENSIONAL ANALYSIS

We have some fractions that are the conversions used to work problems. The next thing to do would be to work some problems, but nearly everyone has trouble with word problems.

Word problems actually tell you how to work themselves. Word problems give the answer in the problem and tell you the information necessary to get there (arrive at the answer).

Example 1.1: A half gallon of milk costs $1.59. How much would 5 gallons cost?

Solution: The answer to the problem is "$." The information necessary to get there is "half gallon of milk costs $1.59" and "5 gallons."

Let's look at the problem again:

problem information	answer	prob. info.
A half gallon of milk costs $1.59.	How much would	5 gallons cost?

The trick to using this information is to write the problem so that it *solves itself* in terms of the units—leave out the numbers for now. The idea is that the problem can be solved in units without the distraction of having the numbers also present. Once the problem is solved in units, then we can put in the numbers and come up with the answer's number.

The answer *in units only* is "$."

$$= \frac{\text{answer}}{\$}$$

To have "$" in the answer, we have to have "$" on the left of the equation. The only "$" we have is "one-half gallon costs $1.59"; we could use that restated in the form of a fraction: "$1.59/half gallon."

$$\frac{\overset{\text{problem}}{\$}}{\underset{\text{info.}}{\text{half gal}}} = \frac{\text{answer}}{\$}$$

"Half gal" is a unit we need to eliminate. The way to do that is to multiply by some value that would put "half gal" in the top of a fraction. Since we have to convert from half gallons to gallons anyway, let's use gallons for the bottom.

$$\frac{\overset{}{\text{half gal}}}{\underset{\textit{conversion}}{\text{gallons}}} \times \frac{\overset{\text{problem}}{\$}}{\underset{\text{info.}}{\text{half gal}}} = \frac{\text{answer}}{\$}$$

Now we have to get rid of "gallons" and we can because half gal/gal is a conversion that we can multiply by the "gallons" we were sent out to buy.

$$\underset{\text{info.}}{\overset{\text{problem}}{\text{gallons}}} \times \frac{\overset{}{\text{half gal}}}{\underset{\textit{conversion}}{\text{gallons}}} \times \frac{\overset{\text{problem}}{\$}}{\underset{\text{info.}}{\text{half gal}}} = \frac{\text{answer}}{\$}$$

Since the only units left are "$" on the left and "$" on the right, the problem is *solved*. All we need to do is put in the numbers and pick up a calculator or pencil and paper.

$$5 \text{ gallons} \times \frac{2 \text{ half gal}}{1 \text{ gallons}} \times \frac{\$ \ 1.59}{\text{half gal}} = \$15.90$$

Solution Summary: A half gallon of milk costs $1.59. How much would 5 gallons cost?

The answer **in units only** is "$."

$$= \frac{\$}{\underline{\quad}}$$

"One-half gallon costs $1.59." We place "$__/half gallon" on the left of the equation because we must have "$" on the left so that $ is equal to $ eventually.

$$\frac{\$}{\text{half gal}} = \frac{\$}{}$$

We convert from half gallons to gallons.

$$\frac{\text{half gal}}{\text{gallons}} \times \frac{\$}{\text{half gal}} = \frac{\$}{}$$

We multiply by the "gallons" we were sent out to buy.

$$\text{gallons} \times \frac{\text{half gal}}{\text{gallons}} \times \frac{\$}{\text{half gal}} = \frac{\$}{}$$

We substitute in the numbers given in the problem statement and solve.

$$5 \text{ gallons} \times \frac{2 \text{ half gal}}{1 \text{ gallons}} \times \frac{\$ 1.59}{\text{half gal}} = \$15.90$$

The above method of solving problems is called dimensional analysis—we work with the "dimensions," the units, to solve the problem. The steps we use are:

1. Identify the answer (units only) from the word problem.
2. Place the answer (units only) after the "=" and the other problem information on the other side of the "=".
3. Get rid of all units, except those on the answer. We do this by multiplying fractions and canceling out everything we don't want.
4. Once the problem is solved in units (the left of the "=" is the same as the right), plug in the numbers from the problem and the conversions used.
5. Calculate the answer.
6. Express the answer to either the correct number of significant digits (later in this chapter) or to a logical answer. The choice of a logical answer is due to what you are calculating; for instance, money answers would be in cents (2 digits past the decimal point).

DISTANCE

The distance is any measurement between two points. You could measure the length of a line from New York City to San Francisco, or maybe the line from the soles of your feet to the top of your head, or, maybe, the line through the center of the Earth.

If we are to use inches as the basic unit of the English system, any distance can be calculated based on inches—the answer may be a very large or small number but can be expressed in inches. The same is true for metric distances—expression of distances can be measured in centimeters.

Since we have inches to centimeters and the inverse (centimeters to inches) conversions, it should be easy to convert between the two systems. All that we need to do is express whatever measurement is to be converted in terms of the basic conversion factor, 1 in/2.54 cm OR 2.54 cm/1 in, and multiply by which version of the conversion factor will do the job. We have to be careful to multiply so unwanted units will cancel out of the problem, leaving only the units of the answer.

Example 1.2: A person's height is measured as 5' 10". Express this height in meters.

Solution: The plan for solving this problem is that the 5'10" can be converted to inches. Once we are in inches, we can get into the metric system to centimeters and then to meters.

Since there are 12 inches in one foot, 5 feet is equal to 60 inches. Adding the 10 inches, we have 70 inches for this person's height.

Solving the conversion using dimensional analysis looks like this:

$$= \quad \text{in}$$

$$\frac{\text{in}}{\text{ft}} \times \text{ft} + \quad \text{in} = \quad \text{in}$$

$$\frac{12\ \text{in}}{1\ \text{ft}} \times 5\ \text{ft} + 10\ \text{in} = 70\ \text{in}$$

NOTE: Multiplication and division are done before addition and subtraction; therefore, 12 × 5 is done first, then added to 10. A way of avoiding error is to write the term as (12 × 5) + 10.

This problem now becomes a conversion from inches to meters.

$$\text{in} = \quad \text{m}$$

We get rid of the inches by multiplying by a fraction with inches in the bottom, but all fractions have to have a top—how about using centimeters:

$$\frac{\text{cm}}{\text{in}} \times \quad \text{in} = \quad \text{m}$$

We eliminate "cm" by dividing from the bottom of a conversion factor (fraction). Since we want to have the answer in meters, we could put "m" in the top of the fraction.

$$\frac{\text{m}}{\text{cm}} \times \quad \frac{\text{cm}}{\text{in}} \times \quad \text{in} = \quad \text{m}$$

Now we have meters equal to meters and the problem is solved, so let's put the numbers in and get the answer.

$$\frac{1\ \text{m}}{100\ \text{cm}} \times \frac{2.54\ \text{cm}}{1\ \text{in}} \times 70\ \text{in} = 1.78\ \text{m}$$

Solution Summary: A person's height is measured as 5' 10". Express this height in meters.

We express the height in feet and inches in inches only.

$$= \quad \text{in}$$

$$\frac{\text{in}}{\text{ft}} \times \text{ft} + \quad \text{in} = \quad \text{in}$$

$$\frac{12\ \text{in}}{1\ \text{ft}} \times 5\ \text{ft} + 10\ \text{in} = 70\ \text{in}$$

The problem becomes a conversion from inches to meters.

$$\text{in} = \quad \text{m}$$

We multiply by the coversion factor that leads us from the metric system to the English system.

$$\frac{\text{cm}}{\text{in}} \times \quad \text{in} = \quad \text{m}$$

We eliminate "cm" and move to meters.

$$\frac{\text{m}}{\text{cm}} \times \frac{\text{cm}}{\text{in}} \times \text{in} = \text{m}$$

The problem is solved in units. We solve in numbers.

$$\frac{1 \text{ m}}{100 \text{ cm}} + \frac{2.54 \text{ cm}}{1 \text{ in}} \times 70 \text{ in} = 1.78 \text{ m}$$

Example 1.3: One atmosphere of pressure is considered to be 760 mm of mercury. How many inches high is that column of mercury?

Solution: We see that the answer to the problem is "inches," so

$$= \text{in}$$

The problem tells us that the information to be changed to inches is "millimeters."

$$\text{mm} = \text{in}$$

We need to get rid of **mm**. We can multiply by a fraction with **mm** in the bottom. But, we have to put in a top for the fraction—since **mm** and **cm** are both metric, we can put in cm/mm and cancel.

$$\frac{\text{cm}}{\text{mm}} \times \text{mm} = \text{in}$$

Since the units on the left are now in "cm," we can get rid of the "cm" and go directly to **in**, the answer.

$$\frac{\text{in}}{\text{cm}} \times \frac{\text{cm}}{\text{mm}} \times \text{mm} = \text{in}$$

The units on the left and the right of the equal sign are the same, and the problem is solved. We put in the numbers from the problem and the conversion and calculate the numerical answer.

$$\frac{1 \text{ in}}{2.54 \text{ cm}} \times \frac{1 \text{ cm}}{10 \text{ mm}} \times 760 \text{ mm} = 29.92 \text{ in}$$

The calculator answer for the problem is 29.92126 inches. This answer should be rounded off. Most of the time, pressures are given in two decimal places; therefore, this answer becomes 29.92 inches of mercury.

Alternate solution: Many people prefer to work conversions within the metric system through the main unit of measure—the meter, for distance. The following is a solution through the meter:

$$\frac{1 \text{ in}}{2.54 \text{ cm}} \times \frac{100 \text{ cm}}{1 \text{ m}} \times \frac{1 \text{ m}}{1000 \text{ mm}} \times 760 \text{ mm} = 29.92 \text{ in}$$

AREA

Area problems are solved in much the same manner as distance calculations. The only consideration is that distance only has one dimension, whereas area has two. Recall that calculations of area are, basically, the length (L) multiplied by the width (W).

$$A = L \times W$$

The units associated with the answer in area would be in square units (in², cm², mi², km², etc.).

Suppose you are presented with a square that is 1 inch to a side (Figure 1.2). If you are asked to calculate the area in inches, you will multiply the length, 1 inch, by the width, 1 inch. The answer will be 1^2 in², which is 1 square inch.

To calculate the area of that same object in square centimeters, you multiply the height, 2.54 cm, by the width, 2.54 cm. The answer will be 2.54^2 cm², which is 6.45 square centimeters (cm²). Using this approach, the conversion factor from square inches to square centimeters, without doing the multiplication, is

$$1^2 \text{ in}^2 = 2.54^2 \text{ cm}^2$$

Therefore, the conversion factor for area, which we already know from the distance factor, can be expressed as fractions.

$$\frac{1^2 \text{ in}^2}{2.54^2 \text{ cm}^2} \quad \text{OR} \quad \frac{2.54^2 \text{ cm}^2}{1^2 \text{ in}^2}$$

Just remember that *you must square all the numbers in the conversion factor,* since the units are squared.

Example 1.4: A commercially available sterile gauze square measures 3 inches × 3 inches. If you are looking for that size square in the metric size, what will be the label in square centimeters on the box?

Solution: The answer is going to be in **cm²**; we can place the answer and the problem information.

$$\text{in}^2 = \text{cm}^3$$

We can get rid of the in² by multiplying by a fraction with in² in the bottom. Since we have to have something in the top of the fraction, why not put cm² (our conversion)?

$$\frac{\text{cm}^2}{\text{in}^2} \quad \times \quad \text{in}^2 = \text{cm}^2$$

The units work and now is the time to put in the numbers. We didn't do anything with the original measurement of 3 inches × 3 inches (3×3 in²). Now is the time since the in² has to be a number that is a square (two numbers multiplied together).

$$\frac{2.54 \text{ cm}^2}{1^2 \text{ in}^2} \times 3 \times 3 \text{ in}^2 = 58 \text{ cm}^2$$

The calculated answer is 58.06 cm²; but, the marketing people would probably prefer to use the whole number for convenience on the label, 58 cm² pad.

VOLUME

Volume problems can be handled in the same way as area except that volume is cubic (three numbers multiplied). This means that cubing the basic distance conversion will give you information about volume, just as squaring gave information about area. In other words, the formula for calculating volume

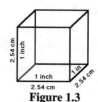

Figure 1.3
One-inch cube.

is length (*L*) times width (*W*) times height (*H*):
$$V = L \times W \times H$$
The volume of the cube in Figure 1.3 would be
$$1 \text{ inch} \times 1 \text{ inch} \times 1 \text{ inch} = 1^3 \text{ inch}^3$$
The reason for the 1^3 is that the three ones are multiplied together. Of course, 1^3 is 1, but there is value in thinking about the conversion including 1^3:
$$1^3 \text{ inch}^3 = 2.54^3 \text{ cm}^3$$
The conversion choices are

$$\frac{1^3 \text{ in}^3}{2.54^3 \text{ cm}^3} \quad \text{OR} \quad \frac{2.54^3 \text{ cm}^3}{1^3 \text{ in}^3}$$

Just remember to *cube everything in the conversion*, just as you squared everything in the area conversion.

Example 1.5: How many cubic meters of air will there be in a room that measures 16 feet long, 12 feet wide, with an 8 foot ceiling?

Solution: The volume of a room is calculated by
$$\text{length} \times \text{width} \times \text{height } = \text{volume}$$
The answer given by the problem is to be m^3. The given information to get there is
$$12 \text{ ft} \times 16 \text{ ft} \times 8 \text{ ft} = \text{ft}^3$$
So the three dimensions of the rectangular cube may be considered as the first piece of information for the problem setup.
$$\text{ft}^3 = \text{m}^3$$
We can cancel out the ft^3 easily using a fraction with ft^3 in the bottom. But fractions have to have a top. Suppose we use inches for the top so we can use the inches to centimeters (but cubed) conversion to get into the metric system.
$$\frac{\text{in}^3}{\text{ft}^3} \times \text{ft}^3 = \text{m}^3$$

We need to cancel inches and can go to centimeters and be in the metric system.
$$\frac{\text{cm}^3}{\text{in}^3} \times \frac{\text{in}^3}{\text{ft}^3} \times \text{ft}^3 = \text{m}^3$$
We can go directly to meters from centimeters and will have the solution.
$$\frac{\text{m}^3}{\text{cm}^3} \times \frac{\text{cm}^3}{\text{in}^3} \times \frac{\text{in}^3}{\text{ft}^3} \times \text{ft}^3 = \text{m}^3$$
This problem is solved with cubic meters on both the right and left of the equals sign. However, we must be careful to plug in numbers that will be cubes for each of the units that are cubed ($12 \times 16 \times 8$ is a cube).
$$\frac{1^3 \text{ m}^3}{100^3 \text{ cm}^3} \times \frac{2.54^3 \text{ cm}^3}{1^3 \text{ in}^3} \times \frac{12^3 \text{ in}^3}{1^3 \text{ ft}^3} \times 12 \times 16 \times 8 \text{ ft}^3 = 43.5 \text{ m}^3$$

Another conversion that is commonly used when dealing with volume is
$$1 \text{ L} = 1.06 \text{ qts}$$
This conversion tells us that the liter is nearly the same size as the quart. It is easy to make a "guesstimate" as to the answer before putting the numbers through a

calculator. Example 1.6 uses this concept to predict the answer.

Example 1.6: We may assume that the average adult has 8 qt of blood for a blood volume. Since blood in the hospital setting is dispensed in liters, how many liters should be ordered for a complete transfusion (complete replacement)?

Solution: The conversion is to be from quarts to liters.

$$qt = \quad L$$

We do not want to keep the **quarts** given to us in the problem statement, so we cancel and, surprise, we can go directly to liters to solve.

$$\frac{L}{qt} \times \; qt = \quad L$$

Notice that a quart and a liter are very close to the numerical value; therefore, the guesstimate is that the answer is close to 8! We substitute in the numbers.

$$\frac{1\,L}{1.06\,qt} \times 8\,qt = 7.55\,L$$

But hospitals do not dispense blood in exact volumes—we would probably have to order 8 liters.

Example 1.7: What is the number of mL per ounce in liquid measurements?

Solution: The answer to this problem is to be in **mL** and the information we have to work with is in ounces (oz). The problem setup would be an equality.

$$oz = \quad mL$$

We do need to get rid of the **oz** because one **oz** is not the same as one **mL**. We can go to quarts from ounces because that will lead us to the conversion between systems (1.06 qt = 1 L) and then go directly to mL.

$$\frac{mL}{L} \times \frac{L}{qt} \times \frac{qt}{oz} \times \; oz = \quad mL$$

We have a solved problem: mL = mL. Placing the numbers and running them through a calculator completes the operation.

$$\frac{1000\;mL}{1\,L} \times \frac{1\,L}{1.06\,qt} \times \frac{1\,qt}{32\,oz} \times 1\,oz = 29.5\;mL$$

The calculated answer is 29.481, which is rounded to 29.5.

> *NOTE: The easiest way to put this through a calculator is to do the bottom first and put it into calculator memory (or write it down if there is no memory available). Then multiply the tops together, press "equals," and then divide by "recall memory."*

WEIGHT

When we perform weight conversions, all we need to do is use an appropriate conversion factor and eliminate units we do not wish to keep. There are a number of conversions that may be used; however, one of the easiest to use for most problems is

$$1\,kg = 2.2\,lb$$

Example 1.8: How would the weight of a 185-lb man be recorded in his records in the metric system?

Solution: We must define the answer. This problem calls for a conversion to the metric system—grams, centigrams, kilograms, whatever. Most of the doctors' offices and hospitals request the weight of a person in kilograms. Then the answer to the problem is to be in kilograms.

Additionally, we have to identify the problem statement—what we are given to achieve the answer. In this problem, that is pounds (lb). So

$$lb = \qquad kg$$

The next step, as with the other problems we have solved, is to get rid of the "lb" and work toward "kg." Since the conversion is lb to kg, we set it up as follows:

$$\frac{kg}{\cancel{lb}} \times \qquad \cancel{lb} = \qquad kg$$

This problem is solved because kilograms are equal to kilograms. All we have to do is plug in the numbers and use a calculator. But, suppose you wanted to do a guesstimate; a way would be to realize that a kilogram is little more than twice a pound and divide the number of pounds by 2: $185/2 = \pm92$. Let's see how close this is.

$$\frac{1 \ kg}{2.2 \ \cancel{lb}} \times 185 \ \cancel{lb} = 84.1 \ kg$$

The guess of 92 kg is not far off *and* does give us a general idea of what the answer is.

By the way, the calculated answer is 84.090909. Rounding off to one decimal place is commonly accepted for the medical recordkeeping.

DENSITY

Density is a calculated value that is very specific for the material that is being considered. Since density is very exact for both pure elements and compounds, the identification of materials can be aided by the measurement of density and comparison against the known values. This is like a detective collecting a clue to solve the mystery of the composition of an unknown.

Density measurements are used in many diverse fields, including geology, engineering, chemistry, and medicine. Density measurements may be used in quality control of manufactured products.

One value of density is that a measurement that is not within normal ranges indicates that there may be something out of the ordinary. Suppose the density of blood is less than the norm, then the water content of the blood is too high. This information, along with other symptoms (clues), helps diagnosis so the treatment will be effective.

The calculation of density is by the following formula:

$$density = \frac{mass}{volume}$$

The units of density are normally expressed in the metric system. Mass is in grams and volume is measured in cubic centimeters. Occasionally, g/cm^3 does not fit the situation, such as when you deal with gases—grams per liter could be used as the unit (gases are much less dense than either liquids or solids). The use of mg/cm^3

or g/L might make more sense.

Example 1.9: A block of a metal is 5.0 cm long by 3.0 cm wide and 4.0 cm high. The block weighs 500 g. What is the density of the block of metal?

Solution: This problem type depends on a formula. The key to working these problems is, of course, to know the formula. It is always a good idea to write out the formula to be used and to substitute in the information given in the problem.

$$D = \frac{mass \ (g)}{volume \ (cm^3)}$$

Substituting in the mass is from the problem statement (500 g), but the volume is not given directly. We have to get volume into the formula, and that can be done if we look at volume closely:

volume = length × width × height

"length × width × height" can be substituted in for "volume"

$$D = \frac{mass \ (g)}{L \times W \times H \ (cm^3)}$$

What is interesting about a mathematical formula is that *it is a set of instructions telling how to work the problem.* All that we have done is to follow the instructions and get to the point where we can put in the numbers from the problem statement.

$$D = \frac{500 \ g}{5.0 \ cm \ \times \ 3.0 \ cm \ \times \ 4.0 \ cm}$$

Since cm × cm × cm is cm³, we have the correct form for the answer (g/cm³).

$$D = \frac{500 \ g}{60 \ cm^3} = \frac{8.33 \ g}{cm^3}$$

The interpretation of the answer is that 1 cm³ of the metal weighs 8.33 grams.

Example 1.10: A 45-gram sample of a liquid occupies a volume of 53 mL. What is the density of the liquid?

Solution: The density formula does not contain mL. There is no problem because we don't really have something new here. *The volume of 1 mL occupies the volume of 1 cm³*; this is a handy conversion within the metric system.

$$\frac{mass \ (g)}{volume \ (cm^3)} \quad OR \quad \frac{mass \ (g)}{volume \ (mL)}$$

We may substitute the problem information directly into the formula.

$$D = \frac{mass \ (g)}{volume \ (mL)} = \frac{45 \ g}{53 \ mL} = \frac{0.85 \ g}{mL}$$

SPECIFIC GRAVITY

Specific gravity is the comparison of the density of a sample to a standard (known material), more often than not water. Water is easy to obtain in the pure state (distilled) and does not respond to most environmental conditions. The density of water is pretty stable at approximately 1 g/cm³ (1 g/mL).

The formula for specific gravity is

$$SG = \frac{\text{density of the sample } (g/mL)}{\text{density of the standard } (g/mL)}$$

Since g/mL is to be divided by g/mL, there are no units on the answer to specific gravity problems.

Example 1.11: The specific gravity of a sample is 0.875. Find the weight of the sample if it measures 15.00 cm × 5.00 cm × 3.00 cm.

Solution: We are not told the volume, but, this is no problem since $L \times W \times H$ is substituted for density. We are not told the density of the standard—this is not a problem, either—the standard is water unless we are told differently by the problem.

Let's start by writing the general formula for specific gravity.

$$SG = \frac{\text{density of the sample } (g/mL)}{\text{density of the standard } (g/mL)}$$

The next thing to do is to get the volume information into the formula for the sample and the density information of water. Suppose we don't use the traditional x; let's use × for multiplication. Since we don't know the weight, let's use w for the unknown.

$$0.875 = \frac{{}^{w}\!/_{15.00 \times 5.00 \times 3.00 \text{ cm}^3}}{1.00 \text{ } {}^{g}\!/_{cm^3}}$$

We now have a **RATIO AND PROPORTION**, two fractions equal to each other. Even though the 0.875 doesn't look like a fraction, we can make it look like one by placing it over 1. We may take any number or symbol and divide by 1 to produce a fraction; there is no change in the value of the original number.

$$\frac{0.875}{1} = \frac{{}^{w}\!/_{15.00 \times 5.00 \times 3.00 \text{ cm}^3}}{1.00 \text{ } {}^{g}\!/_{cm^3}}$$

To solve for w we can cross multiply to get rid of the fractions.

$$\frac{0.875}{1} \diagtimes \frac{{}^{w}\!/_{15.00 \times 5.00 \times 3.00 \text{ cm}^3}}{1.00 \text{ } {}^{g}\!/_{cm^3}}$$

The equation becomes

$$\frac{0.875 \times 1.00 \text{ } g/cm^3}{1} = \frac{1 \times w}{15.00 \times 5.00 \times 3.00 \text{ cm}^3}$$

The right side of the equation is a fraction! We rewrite it so that we have another ratio and proportion by placing the left side over **1**. We can cross multiply again.

$$\frac{0.875 \times 1.00 \text{ } g/cm^3}{1} \diagtimes \frac{1 \times w}{15.00 \times 5.00 \times 3.00 \text{ cm}^3}$$

The cross multiplication gives us a linear setup (no fractions).

$$w = 0.875 \times 1.00 \text{ } g/cm^3 \times 15.00 \times 5.00 \times 3.00 \text{ cm}^3$$

Let's look at the units before we go any further. The value of w should be in weight units, grams in this problem. We find that g/cm³ is to be multiplied by cm³— cm³ is going to cancel. That leaves us with grams—this problem has been solved correctly!

$$w = 0.875 \times 1.000 \text{ g/cm}^3 \times 15.0 \times 5.0 \times 3.0 \text{ cm}^3 = 197 \text{ g}$$

Solution Summary: The specific gravity of a sample is 0.875. Find the weight of the sample if it measures 15.00 cm $\times$ 5.00 cm $\times$ 3.00 cm.

$$\frac{0.875}{1} \bowtie \frac{w/15.00 \times 5.00 \times 3.00 \text{ cm}^3}{1.00 \text{ g/cm}^3}$$

$$\frac{0.875 \times 1.00 \text{ g / cm}^3}{1} \bowtie \frac{1 \times w}{15.00 \times 5.00 \times 3.00 \text{ cm}^3}$$

$$w = 0.875 \times 1.000 \text{ g/cm}^3 \times 15.0 \times 5.0 \times 3.0 \text{ cm}^3 = 197 \text{ g}$$

A note about Example 1.11: Cross multiplication is a quick and easy method of moving elements of the equation from one side to another.

If we had to solve for *A* from the equation

$$\frac{A}{B} = \frac{C}{D}$$

we could multiply both sides by *B* to move it to the other side of the equals sign and isolate *A*.

$$B \times \frac{A}{B} = \frac{C}{D} \times B$$

The *B*'s on the left cancel out because *B* divided by *B* is 1.

$$A = \frac{CB}{D}$$

Cross multiplication would have produced the same results with a lot less stress on you, especially as the fractions become more complex. The same equation solved by cross multiplication is as follows:

$$\frac{A}{B} \bowtie \frac{C}{D}$$

Cross multiplication yields

$$AD = BC$$

Then, dividing both sides by *D* would isolate *A*.

$$\frac{AD}{D} = \frac{BC}{D}$$

Since *D/D* on the left equals 1,

$$A = \frac{BC}{D}$$

TEMPERATURE SCALES

In the Fahrenheit scale, the normal freezing point of water is 32°F and the boiling point is 212°F whereas the Celsius (Centigrade) scale uses 0°C for the freezing point and 100°C for the boiling point. In most countries, the Celsius scale is used. Notice that a 100-point scale (another simplification) is used by the Celsius scale and that a 212-point scale is used by Fahrenheit.

The conversion between the scales is
$$°F = 1.8 \times °C + 32$$
and we can convert either from °F to °C or from °C to °F by substituting into the formula and solving for the answer. One thing to keep in mind is that °F are smaller than °C—that way we can do a quick "check" on any conversion.

Example 1.12: If the temperature is 25°C, what is that temperature in °F?

Solution: This problem requires a mathematical formula. Let's write the formula.
$$°F = 1.8 \times °C + 32$$
Substituting in the problem statement values and remembering that multiplications and divisions are performed before additions and subtractions.
$$°F = 1.8 \times 25 + 32$$

$$°F = 45 + 32 = 77$$
The answer makes sense because the Fahrenheit answer above 0° should be higher than the Centigrade number.

Example 1.13: It is 72°F outside. Express that temperature in degrees Celsius?

Solution: We need to substitute into the formula and solve for the answer since the formula is not in °C.
$$°F = 1.8 \times °C + 32$$

$$72°F = 1.8 \times °C + 32$$
The first step is to get the 32 moved to the other side. This is done by subtracting from the right, but whatever is done to one side of an equation *must* be done to the other.
$$72°F = 1.8 \times °C + \cancel{32}$$
$$-32 \qquad\qquad -\cancel{32}$$
Since $32 - 32 = 0$, the 32 has been eliminated from the right side of the equation. Therefore,
$$40 = 1.8 \times °C$$
Dividing both sides by 1.8 will get the °C to stand alone because $1.8/1.8 = 1$, giving us 1°C. Normally, the 1 is not written when multiplied by a symbol.
$$\frac{40}{1.8} = \frac{1.8}{1.8} \times °C$$
Performing the calculations provides us with the answer.
$$°C = 22$$

Another temperature scale that is often used is the Kelvin scale. The Kelvin scale has 0 as absolute zero. The divisions on a Kelvin thermometer are the same size as a Celsius thermometer. Absolute zero is considered to be –273°C (actually, the value is –273.15°C). If we use the Kelvin scale, there are no negative temperatures, only positive. The conversion is
$$K = °C + 273$$
Notice that the degree sign (°) is traditionally left out when discussing the Kelvin scale.

Example 1.14: What is room temperature in the Kelvin scale?

Solution: Room temperature is considered to be 25°C. It is good form to write the general equation and then perform substitutions.

$$K = °C + 273$$

Plug in the problem information and calculate the answer.

$$K = 25°C + 273 = 298$$

ROUNDING OFF AN ANSWER

When you run a problem through a calculator or do it by hand, the answer usually doesn't come out to be a whole number. The very last step in working a problem is to round off the answer. There are two ways of rounding off so the answer does not go on forever.

The first technique is to round off to some number that is logically based on the use of the answer. For instance, if you were calculating the size of a car's engine in cubic inches from liters, the answer that car manufacturer would probably use the whole number, if at all possible. But, if you were converting someone's height in inches to centimeters for a doctor's records, the chances are that the answer would be expressed to the nearest tenth (0.1) of a centimeter. Of course, the accuracy would actually be determined by the policies of the doctor's office.

To round off simply, the rules are as follows:

1. If the part to be rounded off is less than a 5, drop it.

32.3_499_ becomes 32.3

25.0_6_ becomes 25.0

0.001_19_ becomes 0.001

2. If the part to be rounded off is a 5 or more, increase the number to be held by 1.

177.4_500_ becomes 177.5

0.8_85_ becomes 0.9

4_89_ becomes 500

There are more complex rules for rounding off, but this simple approach is sufficient for the vast majority of problems.

The second technique for rounding off is using the appropriate number of **SIGNIFICANT DIGITS** (SDS).

This technique makes allowances for error in the measurements made. Suppose you measure someone's height in centimeters using a ruler marked in tenths (0.1) of a centimeter. The person is 180.6 centimeters. How accurate is that? Since you can only read to the nearest tenth of a centimeter, the answer can be as low as 180.55 or as high as 180.64.

The number of significant digits in a number does not include place holders— 0._1_ has one SD; 0.00_5_ has one; _503_ has three; _100.100_ has six because the last "0" indicates that the number is accurate plus or minus one half (up or down) of the next place.

The use of significant digits expresses the accuracy of an answer. The number of digits used is from the least accurate number (*the least number of significant digits*).

So $5.1 \times 6.00 = 30.6$ becomes 31 because 5.0 has two SDs and 6.00 has three, and the answer should be in two SDs.

QUESTIONS AND PROBLEMS

1. Express the following measurements in the final units requested.
   (a) 1 m = _____ mm   (b) 2.5 mm = _____ cm
   (c) 0.32 km = _____ m   (d) 5,926 Å = ___ m   [1 Å = 0.00000001 cm]
   (e) 15 mL = _____ L   (f) 156 L = _____ ML
   (g) 0.0062 mL = _____ µL   (h) 9 cm^3 = _____ mL

2. A patient is 5 feet, 6 inches tall. Express in centimeters.

3. You note that the diameter of a patient's pupil is $\frac{1}{8}$ inch. How would this measurement be expressed in centimeters?

4. What is the distance of 32 feet and 9 inches in (a) cm? (b) dm? (c) m?

5. A boat is listed in the sales sheet as being 16 feet long.
   (a) How long is the boat in centimeters? (b) meters? (c) kilometers?

6. A desk is 40 inches wide. What is that width in (a) centimeters? (b) meters?

7. A work space is 60 inches wide and 20 inches deep. What is the area of that work space in (a) square inches? (b) square feet? (c) square yards? (d) square centimeters? (e) square meters?

8. Suppose you were in Europe, where nearly everything is measured in the metric system, and you wanted to buy 4×4 inch gauze pads. What would the label on the box read (whole numbers) if you were looking for those dimensions in centimeters?

9. A room's dimensions are 11 feet by 16 feet. What is the floor space of that room as expressed in square meters?

10. How many square meters of carpet would be necessary to carpet a room that measures 14 feet by 22 feet?

11. A piece of filter paper is $4\frac{1}{2}$ inches in diameter. Assuming that the area is calculated by $A = \pi r^2$, what is the area of that filter paper in square centimeters?

12. During a skin graft, a surgeon is required to move a 17.3 cm^2 piece of skin from one location to another. How many square inches of skin is that?

13. One of the processes necessary in the developing of X-ray film is a soak in a solution called the fixer. The fixer preserves the X-ray film and keeps it from fading. The instructions call for you to replace the fixer after 100 8×10 sheets of film per gallon. How many 5×7 X-ray films could be fixed by one gallon?

14. You pull up to a gasoline pump that charges $0.35 per liter. What would be the price of a gallon of gasoline?

15. The average blood volume of an adult is in the range of 8.0 quarts. Express that blood volume in liters.

16. Assume that you are given the choice of buying one liter of a liquid refreshment and one quart of the same refreshing liquid for the same price. Which is the better buy and by what percentage?

17. A standard drinking glass size is 14 ounces. Express that volume in liters.

18. How many ounces is 250 mL?

19. What is the volume (in cubic meters) of a room that measures 10 feet by 14 feet and has a 7-foot ceiling?

20. The weight of a person is measured to be 190 pounds. Express in kilograms.

21. Many medications are to be calculated in terms of kilograms body weight. Suppose a medication calls for 6 mg/kg body weight.
    (a)   What will be the weight of the medication for a 190-pound patient?
    (b)   Suppose you had calculated the weight of the medication per pound *instead of per kilogram*; how much medication would be given?
    (c)   Have you overmedicated or undermedicated the patient and by how much?

22. An automobile weighs 2 tons. What is the weight (a) in kg? (b) in Mg?

23. A liquid medication is to be administered at the rate of 0.10 mL/kg body weight/hour intravenously. How much medication will have to be ordered from the pharmacy if a 175-pound patient is to be medicated for 8 hours?

24. What is the density of a substance if 2 liters has a weight of 2,950 grams?

25. If the density of hydrogen is considered to be 0.08988 mg/mL, what will be the weight of 6 liters of hydrogen?

26. The density of water is usually considered to be 1 g/mL at 15°C. What is the weight of one gallon of water at 15°C?

27. If the density of lead is 11.3 g/cm^3, what is the weight of 1 ft^3 of lead?

28. The density of solid gold is 19.31 g/mL. Suppose you were told that you could have a 5 gallon bucket full of gold *if* you could carry it one-half mile without putting it down. Would that bucket of gold be yours?

29. The density of mercury is 13.59 g/mL. Will a block of lead (D = 11.3 g/mL) float in a container of mercury? Explain.

30. What is the specific gravity if an iceberg floats one-ninth out of water?

31. If the density of blood is 1.045 g/mL, what is the specific gravity of blood as compared to xylene, which has a density of 0.8802 g/mL?

32. To obtain the weight of hydrogen in the laboratory, you would have to weigh it in a vacuum. Why?

33. Your average body temperature is 98.6°F. What is your temperature in °C?

34. If water boils at 100°C, what is the boiling point of water in °F?

35. What is room temperature in °F?

36. What is room temperature in K?

37. An alcohol boils at 150°F. Calculate the boiling point of the alcohol in K.

38. What is the number 360.257 when rounded off to the nearest tenth?

39. Round off 0.004361 to the nearest thousandth.

40. What is the number of significant digits for 32.6 cm × 25.998 cm × 0.012 cm?

ATOMIC STRUCTURE

• Periodic Table Format • Symbol • Atomic Number •
• Atomic Weight • Atomic Parts • Atomic Structure •
• Patterns • Isotopes • Isotope Ratios •
• Questions and Problems •

PERIODIC TABLE FORMAT

The Periodic Table of the Elements (Table for short) is a tool for the chemist and the chemistry student. If the Table is available for reference and you know how to read it, your life as a student of chemistry becomes much easier than if you were to memorize the information the Table contains. Before you read further, please take a look at the Table printed on the inside front cover of this manual and Figure 2.1 to get a feel for it.

The Table is designed to be a logical presentation of the elements. The elements are laid out in order to show similarities and trends in characteristics, such as atomic structure factors, boiling and freezing points, densities, chemical activities, crystalline structures, and the likely oxidation numbers (charges in a compound). The Table may even be used to predict information about elements that is not available.

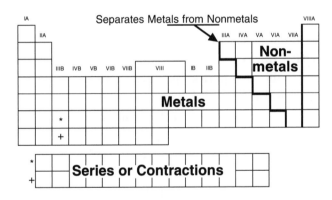

Figure 2.1
The Periodic Table of the Elements.

The Table used in this manual is organized into columns, called **groups**, which are named with a Roman numeral and followed in most cases by a letter (A or B). The rows are called **periods** and have a name that is the number of the period counting down from the top; there are seven periods.

There are two rows at the bottom that are separated from the Table proper. These are called **series** or **contractions**. A series belongs placed on the Table stacked up like a deck of cards as the position indicated in Group IIIA, Period 6 or 7. Of course, displaying these series is a major printing problem; we don't have three-dimensional printing capabilities. The placement of the series in such a tight location is due to the close similarity of characteristics of the elements involved.

The diagonal "staircase" toward the right of the Table separates the metals on the left from the nonmetals on the right. Those elements located along the separation are metalloids—elements that are neither metals nor nonmetals (or may have characteristics of both).

The last major division of the Table is the separation of the last Group, Group VIIIA, from the table proper by a heavy line. This Group may be named Group 0, the Noble Gases, or the Inert Gases. The last name, Inert Gases, tells a story—the elements are gases and are inert. *Inert* is used to indicate that these elements normally do not enter into a chemical compound. Although some of these elements do enter into chemical compounds, their chemistry is outside the scope of this book.

SYMBOL

The entries in the Table are each placed in their own cell or box (Figure 2.2). To avoid confusion, the symbol in the center of the box cannot be repeated. You will notice that each symbol starts with a capital letter and, if there is a second letter, the second letter is lowercase (small letter). If this system is not used carefully, there can be some major problems.

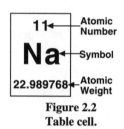

Figure 2.2
Table cell.

Suppose you wanted to write the symbol for carbon monoxide, **CO**, and did so as **Co**. You didn't write the symbol for carbon monoxide; you wrote the symbol for cobalt. This would lead to miscommunication and could create some very real problems.

ATOMIC NUMBER

Each of the cells of the Table (Fig. 2.2) contains a *whole number* at the top called the **ATOMIC NUMBER** (at. no.). All of the atomic numbers may be read from left to right and then down to the next period, just like reading a book. The only breaks in this pattern are at Period 6, Group IIIB and Period 7, Group IIIB. The series (contractions) belong in those positions, so there are no breaks in the pattern, just sidesteps.

ATOMIC WEIGHT

The number at the bottom of the cell (Fig. 2.2) is the **ATOMIC WEIGHT**. The atomic weights are mostly fractional values because they are *average* atomic weights. "Average" atomic weight means that there is more than one form of the element (isotopes), which will be discussed later in this chapter.

Some of the atomic weights are whole numbers, but you will notice that they are in parentheses. The parentheses indicate that there is not enough information available about the element to determine an atomic weight with more accuracy than the whole number.

ATOMIC PARTS

The atomic weight, rounded-off to a whole number, is the *number* of protons (P) and neutrons (N). The atomic number, which is a whole number, is the number of protons.

If we subtract the atomic number (P) from the rounded off atomic weight (P + N), we arrive at the number of neutrons (N).

$$
\begin{array}{lr}
\text{atomic weight} & P + N \\
\underline{-\text{ atomic number}} & \underline{-P} \\
\text{number of neutrons} & N
\end{array}
$$

Of course, it would be nice to have some meaning for this information.

The atom is divided into geographic locations: the nucleus in the center and the orbits (shells) around the nucleus as in Figure 2.3. The nucleus is composed of protons and neutrons.

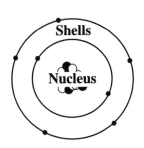

Figure 2.3
The atom.

The proton is a small bit of matter that has a weight of 1 amu (atomic mass unit). The value of an amu is a tiny fraction of a gram; therefore, the amu is really an arbitrary value assigned so that we don't have to refer to an actual value that is so small. The proton has an electrical charge of +1. Again, the value of the charge is very small; the +1 is used for convenience.

The nucleus also contains the neutrons. Each neutron is considered to have a weight of 1 amu . The charge on a neutron is neutral (zero)—easy to pick up from the name.

The shells (orbits) contain the electrons (e^-). Each electron has a charge of –1. This charge is directly opposite of the proton's charge. The mass of an electron is very small, often given as 1/2,000 amu and not really significant when we round off the atomic weight of an atom.

By definition, the charge on an atom is zero. This means that there must be the same number of protons and electrons to get the zero charge. Therefore, the atomic number, the number of protons, is also the number of electrons.

ATOMIC STRUCTURE

Now we can outline the structure of an atom using the atomic number and the rounded-off atomic weight. If we want to determine the structure of sodium (Fig. 2.2), we will place 11 protons in the nucleus because the atomic number is 11. The atomic number minus the atomic weight (23 – 11) tells us the number of protons, 12, in the nucleus. Furthermore, the atomic number indicates that there will be 11 electrons in the shells (orbits).

atomic weight	23	P + N
– atomic number	– 11	P
number of neutrons	12	N

The nucleus contains 11 protons and 12 neutrons; there are 11 electrons in the shells.

Placing the electrons in the shells for the first 20 elements (atomic numbers 1 through 20) depends on knowing the maximum number of electrons that will go into each orbit.

1st shell maximum   = 2 e^-
2nd shell maximum   = 8 e^-
3rd shell maximum   = 8 e^-

Imagine that we are dealing with a fountain like those found in old town squares (Fig. 2.4). When the top bowl is filled, the remainder of the water spills into the second level, which is larger than the first bowl. When full, the water will spill into the third bowl, etc.

In the case of sodium's shells, the first two electrons will go

Figure 2.4
Fountain.

into the first shell filling that shell. The remainder (nine electrons) will flow into the second shell filling the second shell with eight electrons. The remaining electron will flow into the third shell.

A simple way to show this is to indicate the nucleus and contents, and then each shell.

```
        \   \   \
11P
        2   8   1
12N
        /   /   /
```

Example 2.1: What is the atomic structure of chlorine?

Solution: The information we need is directly from the Periodic Table. The rounded off atomic weight is 35; the atomic number is 17. Then

atomic weight	35	P + N
− atomic number	− 17	P
number of neutrons	18	N

Then

```
        \   \   \
17P
        2   8   7
18N
        /   /   /
```

Example 2.2: Give the complete atomic structure of the carbon atom. The Periodic Table of the Elements tells us that carbon is atomic number 6 and has an atomic weight of 12.011.

Solution: We will use the rounded-off atomic weight of 12 and the atomic number (6) to determine the numbers of particles present.

atomic weight	12	P + N
− atomic number	− 6	P
number of neutrons	6	N

Express in terms of a structure.

```
        \   \
6P
        2   4
6N
        /   /
```

NOTE: Those elements beyond atomic number 20 fill their shells by different patterns. The patterns will be covered later.

PATTERNS

We have some information about the structure of elements that appears to fall into a pattern.

Look at the atomic structures of sodium, chlorine, and carbon. Then look at their position in the Table and the number of electrons in the outer shell (orbit) compared to the group number.

Element	e⁻ in Outer Orbit	Group
Sodium	1	IA
Chlorine	7	VIIA
Carbon	4	IVA

The group name, if it is an A group, tells us how many electrons are in the outside shell (orbit) of the first atomic number 20 elements. It turns out that this is true all the way down the A groups, except for helium at the top of Group VIIIA with 2 electrons. Since carbon is in Group IVA and has four electrons in the outside shell, lead (Pb) at the bottom of Group IVA also has four electrons in the outside shell.

If this pattern holds true to determine the number of electrons in the outside shell, then the number of electrons in the outside shell of oxygen (Group VIA) should be six. Let's see. Look up the atomic number and atomic weight; then

$$\begin{matrix} & & \backslash & \backslash \\ 8P & & & \\ & & 2 & 6 \\ 8N & & & \\ & & / & / \end{matrix}$$

So the pattern holds for oxygen and, therefore, for the entire group.

Another pattern is related to the period number:

Element	Orbits	Period
Sodium	3	3
Chlorine	3	3
Carbon	2	2

Sodium and chlorine are both in the third period and both have three orbits. Carbon is in the second period and has two orbits. Further inspection of the Table shows us that the number of orbits is the same as the period number.

Example 2.3: What is the number of shells and the number of electrons in the outside orbit of lithium (at. no. 3), magnesium (at. no. 12), radium (at. no. 88), and astatine (at. no. 85)?

Solution: We can answer this question by making up a chart and filling in the information.

		Outer Shell		
Element	Group	= Electrons	Period	= Shells
Li	IA	1	2	2
Mg	IIA	2	3	3
Ra	IIA	2	7	7
At	VIIA	7	6	6

This is science and appears to be too easy. Of course, there has to be an exception to this nice, neat pattern, and there is. The exception is helium (at. no. 2). The period number is the number of the outside shell (orbit). But the group name, Group VIIIA, does not tell us the number of electrons in the outside shell because there are only two electrons for the atomic structure, atomic number 2. Therefore, the number of electrons in the outside shell of helium is two, not eight.

ISOTOPES

Earlier in this chapter we mentioned that atomic weights are average atomic weights. Hydrogen has the atomic weight of 1.0079, but why?

First, the calculation of an average is done by adding up all of the members of a sample (weights in this case) and dividing by the number of individuals in the sample.

Hydrogen atoms do not all have the same weight (protons + neutrons). There are three different kinds of hydrogen that exist in nature: hydrogen with a weight of 1, hydrogen (deuterium, D) with a weight of 2, and hydrogen (tritium, T) with a weight of 3. Notice that the "deu" in deuterium has the value of 2 and the "tri" in tritium has the value of 3.

If we were to take all of the three types of hydrogen atoms in the universe, add up their weights, and then divide by the number of atoms, we would come up with an average atomic weight of 1.0079.

By the way, the difference in weights is obvious when we look at the statistics for the hydrogens.

Isotope Name	At. No.	At. Wt.	Symbol
Hydrogen	1	1	H
Deuterium	1	2	D
Tritium	1	3	T

All three hydrogens have one proton because they share the atomic number of 1. Hydrogen does not have any neutrons, Deuterium has one neutron, and Tritium has two neutrons. (All we have to do to determine the number of neutrons is subtract the atomic number from the atomic weight.)

Incidentally, the number of protons is what determines the element with which we are working. The number of protons, as we will discuss later, is what determines the chemical activity of an atom (element). This means that hydrogen, deuterium, and tritium would all be expected to react chemically in much the same way.

All three of these forms of hydrogen are called **ISOTOPES** of hydrogen. Isotopes, then, are atoms with different atomic weights having the same atomic number.

The word *isotope* probably brings to mind a radioactive material. That is not necessarily true. In the case of the hydrogens, hydrogen is not radioactive, or we would be in a lot of trouble because we drink a lot of water, H_2O.

However, tritium is a naturally occurring radioactive isotope of hydrogen. To separate the nonradioactive version from the radioactive version of an element, the term **RADIOISOTOPE** is used for radioactive isotopes.

As far as the radioisotopes of hydrogen are concerned, the atomic weight of 1.0079 means that the vast majority of hydrogen is of the non-radioisotope type, hydrogen (atomic weight of 1). However, there is still a very small amount of radioactive hydrogen around. Have you ever heard of "heavy water"?

What would happen if we were to react tritium with oxygen? The results would be water because tritium reacts to form the same compound, water, as hydrogen does. The molecular weight of H_2O is 18 units (1 each for the two hydrogens and 16 for the oxygen). The molecular weight of T_2O is 22 (3 each of the two tritiums and 16 for the oxygen). Therefore, T_2O is heavier than H_2O. Drinking T_2O, which is radioactive, would not be the healthiest thing to do.

Most of the elements do have isotopes, many of which are radioisotopes. None of the atomic weights are whole numbers except the atomic weights in parentheses. Notice that nobelium (atomic number 102) has an atomic weight of 259 in parentheses. The parentheses mean that there is not enough information known to calculate a more accurate atomic weight—the element is very scarce. Many of these elements are man-made, especially those found in the series (contractions).

Most of the elements do have at least one isotope that is a radioisotope, and some of the elements (uranium, radium, thorium, and many others) have only radioisotopes.

Most isotopes don't have specific names as the three isotopes of hydrogen. The naming system is a direct application of the atomic weight to the name. For instance, carbon has two isotopes, carbon-12 and carbon-14 or the symbols C-12 and C-14.

ISOTOPE RATIOS

Most of the elements do have more than one isotope. If we have the atomic weight and the number of the isotopes, we should be able to calculate the ratio of the isotopes.

The general idea is that the atomic weight from the Table is the *average* atomic weight. That weight is 100% of the total atomic weight. All other atomic weights of the isotopes would be percentages of the whole and would be added to get the whole, average weight.

Example 2.4: Suppose we knew that chlorine had two isotopes of atomic weights 35 and 37. The average atomic weight of chlorine is 35.453 amu from the Table. Calculate the ratio of the chlorine isotopes.

Solution: The average atomic weight, 35.453, is the total—100% of the total. The two, true weights are parts of the whole and would add to be the whole. The total is 100% of the average atomic weight (1×35.453).

If one of the parts is a fraction of the chlorine, W, the other part is $1 - W$. The two parts are added to give us the total.
$$W + (1 - W) = 1$$
The atomic weights times their fraction sum to one average atomic weight.
$$35(W) + 37(1 - W) = 1 \times 35.453$$
The multiplication is done first, and then the addition.
$$35W + 37 - 37W = 35.453$$
Let's collect the terms with the W first.
$$-2W + 37 = 35.453$$
We subtract 37 from both sides of the equation.
$$-2W = -1.547$$
This solution, so far, has negative numbers on both sides. We may multiply all units within the equation by –1 to eliminate the negatives and give us positives:
$$2W = 1.547$$
We divide both sides by 2 will isolate the W.
$$W = 0.77$$
So what does all of this mean? In the setup, we allowed the W to represent the part of the chlorine that was of atomic weight 35. The $1 - W$ is the part of the chlorine that is of atomic weight 37 (we subtract W from 1 to get the part of the chlorine that is Cl-37). Therefore,

0.77 is the fraction of chlorine-35

0.23 is the fraction of chlorine-37

Many people would prefer to see the answer in terms of percentage. To convert these answers to percentage, all we have to do is multiply by 100.

77.% Cl-35 and 23% Cl-37

We aren't quite finished, though. The problem asks for the ratio of the two chlorines—*always reread the problem.*

Dividing the larger number (%) by the smaller (%) will give us the answer of some number to 1.

0.77/0.23 = 3.3 (% calculation = same answer)

Therefore, Cl-35 is more than three times more common than Cl-37.

NOTE: Example 2.4's solution is a linear problem in which there is a mix of addition and multiplication. The order for working these kinds of problems is to perform the operation(s) inside the parentheses, if any, first and then work outside. The order for all manipulations is multiplication and/or division first, then addition and/or subtraction.

Example 2.5: In Example 2.4, we found that the ratio of Cl-35 to Cl-37 is essentially 3:1. Let's work backward to determine the average atomic weight of chlorine.

Solution: If we assume that there are only five atoms of chlorine in the entire universe, we can calculate the total weight.

$$3 \text{ atoms of Cl-35 is} \quad 3 \times 35 = 105 \text{ amu}$$
$$1 \text{ atom of Cl-37 is} \quad 1 \times 37 = \underline{37 \text{ amu}}$$
$$\text{total of the chlorine} \qquad 142 \text{ amu}$$

Then we divide the total weight by the number of chlorines in our universe:

142 units/4 chlorines = 35.5 units per Cl

The calculated average atomic weight for chlorine is 35.5 units.

NOTE: Many chemistry instructors will request that you memorize the symbols for the elements. The suggestion is that you concentrate on identifying those elements for which you already know the symbols, and then identify the elements that have symbols relating to the English name. Notice that these symbols are either the first letter of the name or, if there are two letters, the first letter of the name and a letter of an easily heard sound of the name. Example: Magnesium has the symbol of "Mg." The "M" is the first letter of the name and the "g" is easily heard when pronouncing the name.

The last step is to learn the symbols for those elements that do not have recognizable English names. Iron has the symbol of "Fe," which comes from the Latin name ferrum.

QUESTIONS AND PROBLEMS

1.   Where are protons found within any atom?

2.   What is the location of electrons found in an atom?

3.   What is/are the atomic part(s) that contribute to the atomic number?

4.   What is the number of protons in each atom of helium?

5. What is the number of protons in an atom of iron?

6. How do you know how many electrons are in an atom?

7. How many electrons would be found in an atom of uranium (U, at. no. of 92)?

8. Each atom of lead (Pb, at. no. of 82) would be expected to have how many electrons?

9. How do neutrons contribute to the charges within an atom?

10. Barium (Ba, at. no. 56; at. wt. 137) contains how many neutrons?

11. What is the number of protons, electrons and neutrons contained in an atom of magnesium (at. wt. 24)?

12. What is the overall charge on an atom of sodium?

13. What is the number of electrons in the first orbit of lithium?

14. What is the number of electrons in the outside orbit of carbon?

15. How many electrons are in the outside orbit of helium?

16. Give the number of electrons in the outside orbit of neon and argon.

17. In question #14 you were asked for the number of electrons in the outside orbit of carbon. How many electrons are in the outside orbit of lead?

18. What is the number of electrons in the outside orbit of oxygen?

19. What is the number of electrons in the outside orbit of sulfur?

20. Suppose an unknown element were tested and found to contain 79 protons, 79 electrons, and 197 neutrons. What is the name of the element?

21. Which element would display the following statistics: 98 protons, 98 electrons, 154 neutrons?

22. Chlorine has two isotopes of atomic weights 35 and 37. What is the complete electron configuration of these two isotopes?

23. Chemical activity may be predicted by the number of electrons in the outside orbit of an atom. Would the two isotopes of chlorine react differently from each other?

24. What are the names of the two isotopes of chlorine?

25. Carbon displays an average atomic weight of 12.011. What is the most common of the isotopes of carbon, C-12 or C-14?

26. What is the weight of the most common isotope of uranium, at. no. = 238.0289?

27. Calculate the weight of carbon dioxide, CO_2, produced from the most common isotopes.

28. Given the most common isotopes, give the weight of sulfuric sulfuric acid, H_2SO_4.

29. Lithium has two naturally occurring isotopes. What is the abundance of the lithium isotopes, Li-6 and Li-7?

30. What is the percent composition of nitrogen that is N-15 if the only naturally occurring isotopes are N-14 and N-15?

CHEMICAL COMPOUNDS

• Compound Formation • Oxidation Numbers •
• Polyatomic Ions • Questions and Problems •

COMPOUND FORMATION

Writing chemical compounds is one of those topics that most students look at with some desire to be elsewhere. The topic has a reputation of requiring a lot of memorization. Actually, most of the memorization can be avoided by understanding the predictions of the Periodic Table of the Elements (Table). We can develop a prediction technique by looking at how elements form compounds.

Let's analyze table salt, NaCl, from the standpoint of the atomic structure of the sodium and chlorine in the compound. We use the rounded-off atomic weights.

Sodium has the atomic number of 11 and the atomic weight of 23 (refer to Chapter 2). The atomic number tells us how many protons there are (11), and the atomic weight minus the atomic number tells us how many neutrons are present (12). Also, the number of protons and the number of electrons will be the same in an atom (11 of each). The atomic structure of a sodium atom is

$$\begin{array}{c} \text{11P} \\ \text{12N} \end{array} \quad \begin{array}{ccc} \backslash & \backslash & \backslash \\ 2 & 8 & 1 \\ / & / & / \end{array}$$

The Table on the inside front cover provides us with the information for the chlorine atom. Using the same technique to determine chlorine, we find that there are 17 protons, 17 electrons, and 18 neutrons. The structure of the chlorine atom is

$$\begin{array}{c} \text{17P} \\ \text{18N} \end{array} \quad \begin{array}{ccc} \backslash & \backslash & \backslash \\ 2 & 8 & 7 \\ / & / & / \end{array}$$

How do sodium and chlorine atoms interact to produce a compound? To understand the actions involved, we have to look at Group VIIIA on the Table.

Group VIIIA is often called the inert gas group. The elements are gases; they do not normally react. For all practical purposes, they are inert. The outside orbit of all the inert gas group, except helium, contains eight electrons.

Another consideration is that the formation of chemical compounds is by means of what the electrons do when a chemical compound is formed. Electrons may be lost or gained from an atom.

We can put this information together. The relationship between being stable (not reacting chemically) and the outer orbit appears to be related to having the same number of electrons in the outside orbit as one of the inert gases.

The inert gases closest to sodium on the Table (closest in atomic number) are neon (at. no. 10) and argon (at. no. 18). Both have eight electrons in the outside orbit. If sodium could gain or lose enough electrons, its outside orbit would look just like an inert gas and sodium would be stable (probably would not react further).

The question is, "Will sodium gain seven electrons to look like argon or will sodium lose one electron to look like neon?" It is easier from the standpoint of the energy used to lose one electron than it is to gain seven. Therefore, one electron will be lost and sodium's orbits will look like neon's—stable.

$$
\begin{array}{c}
\text{11P} \\
\text{12N}
\end{array}
\quad {}_2 \quad {}^8
$$

The law of conservation of matter states that we cannot create or destroy matter. We appear to have destroyed an electron. Have we? Not really—that electron could be picked up by something that wants electrons. Let's look at chlorine and see what happens.

Chlorine has seven electrons in the outside orbit. It would take less energy for chlorine to attract one electron and therefore look like argon than for chlorine to lose seven electrons and look like neon.

$$
\begin{array}{c}
\text{17P} \\
\text{18N}
\end{array}
\quad 2 \quad 8 \quad 8
$$

AND there is a spare electron from sodium available to be accepted by chlorine.

$$
\text{Na} \quad
\begin{array}{c}
\text{11P} \\
\text{12N}
\end{array}
\quad 2 \quad 8 \quad \textcircled{1} \longrightarrow 7 \quad 8 \quad 2 \quad
\begin{array}{c}
\text{17P} \\
\text{18N}
\end{array}
\quad \text{Cl}
$$

AND chlorine does accept sodium's electron.

$$
\text{Na} \quad
\begin{array}{c}
\text{11P} \\
\text{12N}
\end{array}
\quad 2 \quad 8 \quad 8 \quad 8 \quad 2 \quad
\begin{array}{c}
\text{17P} \\
\text{18N}
\end{array}
\quad \text{Cl}
$$

Both sodium and chlorine now have the inert gas configuration—they both look like inert gases and will not generally gain or lose more electrons.

> *NOTE: Is the preceding structure of chlorine backwards? Not really. Remember that the atom may be imagined as a nucleus surrounded by its orbits, which are represented as circles.*

You may have noticed that the number of neutrons is required to complete the structure of the nucleus, but it has nothing to do with the filling of the orbits (the determination of the **ELECTRON CONFIGURATION**). We could leave out the neutrons and not affect the portions of the atom that we need to work with the formation of chemical compounds, the protons and electrons.

Example 3.1: The formula for magnesium iodide is MgI_2. Explain why there are two iodine atoms for each magnesium.

Solution: Magnesium has the atomic number of 12; therefore, the number of protons is 12. Since there are 12 protons in the nucleus, there are 12 electrons in the orbits of the magnesium atom. The electron configuration is $2e^-$ in the first orbit, $8e^-$ in the second orbit, and 2 electrons in the third orbit. Magnesium would tend to lose 2 electrons to look like neon rather than gain $6e^-$ to look like argon. (Notice that magnesium is in Group IIA. These elements would all have 2e⁻ in the outside orbit.)

Iodine is in Group VIIA. Group VIIA elements all have 7e⁻ in the outside orbit. Each iodine could gain 1e⁻ to have 8e⁻ in the outside orbit.

Since magnesium is giving up 2e⁻ and each iodine can gain 1e⁻, two iodines are required to produce magnesium iodide, MgI_2.

We cannot use the concept of having eight electrons in the outside orbit if the atom is not capable of holding eight electrons. Lithium (at. no. 3) only has three protons and could not be expected to hold enough electrons to have eight in the outside orbit—that would be 10 electrons all together.

Helium is in the inert gas group, Group VIIIA, and has two electrons in the outside orbit (the first orbit).

If lithium were to do whatever is necessary to look like helium, lithium would be stable. Lithium atoms have two electrons in the first orbit and one in the outside orbit. If lithium were to gain one electron from somewhere, it would not be a perfect helium imitator.

If lithium were to lose one electron, it would look just like helium (two e⁻ in the outside orbit—first orbit).

Of course, we would have to account for the lost electron. There would be an atom nearby that would be willing to pick up the electron lithium released. If that atom happened to be chlorine, the compound produced would be LiCl because lithium is giving up one electron and chlorine will gain one electron. In other words, only one chlorine is necessary to satisfy the law of conservation of matter problem generated by lithium's loss of an electron.

Example 3.2: On the basis of the electron configurations, what would be the formula of boron fluoride?

Solution: Boron has the atomic number of 5. The electron configuration of boron is 2e⁻ in the first orbit and 3e⁻ in the second (outside) orbit. Boron would tend to lose three electrons instead of gaining five electrons to achieve an inert gas configuration.

Fluorine's atomic number is 9, indicating 2e⁻ in the first orbit and 7e⁻ in the outside orbit. Fluorine would gain one electron.

Since boron is losing three electrons and fluorine is gaining one electron each, three fluorines are necessary. The formula is BF_3.

Alternate Solution: Boron is in Group IIIA and, therefore, has three electrons in the outside orbit. Boron would be most likely to lose three electrons rather than gain five.

Fluorine is in Group VIIA and has seven electrons in the outside orbit. Fluorine is an ideal candidate to gain one electron.

Since boron will lose 3e⁻ and each fluorine will gain 1e⁻, three fluorines are required for each boron.

The formula of boron fluoride is BF_3.

Note: The number of electrons in the A Group elements is the same as the group number. The exception is helium, Group VIIIA, with two electrons in the outside orbit. This shortcut does not work with the B groups. The transition metals are handled later in this manual.

OXIDATION NUMBERS

If we had to write chemical compounds on the basis of the electron configuration, it would take forever to figure out each one. There is a much quicker way.

Let's take a look at what happened to sodium when we wrote the structure of NaCl. The atom, Na, lost an electron to achieve an inert gas configuration.

$$\begin{array}{llll} 11P & \backslash & \backslash & \backslash \\ & 2 & 8 & 1 \qquad \text{as an atom} \\ 12N & / & / & / \end{array}$$

becomes

$$\begin{array}{lll} 11P & \backslash & \backslash \\ & 2 & 8 \qquad \text{in a compound} \\ 12N & / & / \end{array}$$

The atom has an overall charge of zero. There are the same number of protons (+1 each) as there are electrons (−1 each), which would add up to zero. Sodium in a compound is a different story; there are 11 protons and 10 electrons. Adding the charges (P + e⁻) would give us an overall charge of +1 (N's = 0).

$$\begin{array}{lll} 11P & \backslash & \backslash \\ 12N & 2 & 8 \\ & / & / \end{array}$$

$$+11 \qquad -10 \qquad\qquad +11 - 10 = +1$$

Since all of the members of Group IA have one electron in the outside orbit, they should all behave in the same manner as sodium.

Let's look at a Group IIA element: how about magnesium from Example 3.1. We found that magnesium had two electrons in the outside orbit, which is predictable since magnesium is in Group IIA. Magnesium lost the two electrons from the outside orbit.

$$\begin{array}{lll} 12P & \backslash & \backslash \\ 12N & 2 & 8 \\ & / & / \end{array}$$

$$+12 \qquad -10 \qquad\qquad +12 - 10 = +2$$

Therefore, the overall charge on magnesium in a compound is +2. Neutrons (0) do not contribute to charge.

All of the Group IIA elements are like magnesium; they have two electrons in the outside orbit. They will all have a +2 charge in a compound.

It is beginning to look like there is a pattern—Group IA elements display a charge of +1 and Group IIA elements have a charge of +2 in a compound. Will Group IIIA do the same thing? How about Group VIIA? Let's take a look at Al from Group IIIA and Cl from Group VIA, ignoring the neutrons.

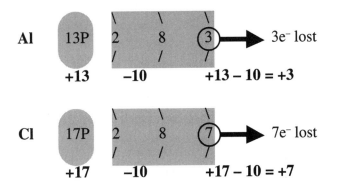

It looks like there is predictive value to the Table in the sense that the A groups will predict the positive overall charge of an element when it enters into a compound. Of course, this does not apply to Group VIIIA because those elements generally do not react (the basis of our original assumption).

Notice that we have been discussing predictions. There are very few times in science that words like *always* and *never* work. There are exceptions to nearly all the rules and laws of science. Figure 3.1 is a summary of the predictions we have been discussing.

By the way, the chemist does not refer to the "overall" charge. The term that is used for this concept is **OXIDATION NUMBER**, which is the charge displayed in a compound. Another name for the oxidation number is **VALENCE**.

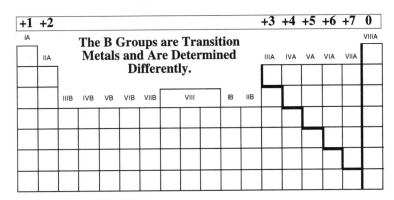

Figure 3.1
Positive oxidation numbers for A groups.

This is fine except that chlorine, as we have seen, can also have an oxidation number of –1. The oxidation number of +7 results in the loss of seven electrons; the oxidation number of –1 occurs when one electron is gained.

What will happen if oxygen (Group VIA) were to gain electrons instead of losing them?

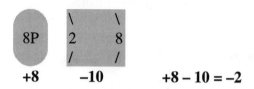

With the gain of the 2e⁻, oxygen becomes

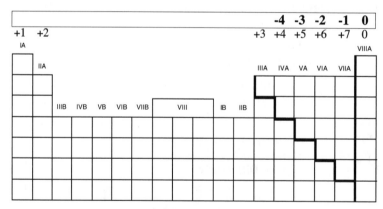

+8   −10   +8 − 10 = −2

Oxygen can have an oxidation number of –2.

We can develop a trend from the Table that will help us with the negative oxidation numbers.

1. Group VIIIA has the oxidation number of zero.
2. Group VIIA (chlorine) can have an oxidation number of –1.
3. Group VIA (oxygen) can be –2.

Then Group VA can be –3 and Group IVA can be –4. Actually, this trend in negative oxidation numbers stops with Group IVA because the remainder of the Periodic Table is mostly metals (refer to the inside front cover of this manual). A chemical characteristic of metals is positive oxidation numbers.

Figure 3.2 includes the predictions of the negative oxidation numbers we have just discussed.

Figure 3.2
Negative oxidation numbers for A groups.

Once an atom has gained or lost electrons, it is no longer considered to be an atom but an **ION**.

We can use ions to write chemical compounds. By agreement among chemists, compounds have a **NET CHARGE** (overall charge) of zero. All we have to do to write a chemical compound's formula is use the oxidation numbers of the ions involved and make certain that the net charge is zero.

There is more than one technique for writing chemical formulas. Take a look at Examples 3.3 and 3.4.

Example 3.3: Determine the formula of boron chloride on the basis of the predicted oxidation numbers from the Periodic Table of the Elements.

Solution: We find boron in Group IIIA, which means that boron is predicted to display an oxidation number of +3.

Chlorine is in Group VIIA. Group VIIA elements may have oxidation numbers of +7 and –1. Chlorine that has a negative charge is renamed *chloride*.

Compounds (by the definition of a compound) have a net charge of zero.

Since boron may only be a positively charged ion, then chlorine *must* be negatively charged. The opposite charges are required so the overall (net) charge will be zero. Since boron is +3 and chloride is -1, three chlorides are required to provide us with a net charge of zero.

$$B^{-3} + Cl^{-1} + Cl^{-1} + Cl^{-1} = \text{zero net charge}$$

Boron chloride has the formula of BCl_3.

SHORTCUT: Notice that boron is a +3 and chloride is a -1. Suppose we just "cross the charges" like this
$$B^{+3} \diagdown Cl^{-1}$$
becomes
$$B_1 Cl_3$$
We don't write ones unless we absolutely have to, so the compound is
$$BCl_3.$$

The shortcut looks great. However, there is one problem, as shown in Example 3.4. The ratio of the number used to write the compound should be the lowest number possible. This means that the common factors must be removed. Let us look at Example 3.4.

Example 3.4: What is the formula of the compound of magnesium and oxygen?

Solution: Using the Table to determine the charges tells us that magnesium (Group IIA) can *only* be a +2. Therefore, oxygen must be a negative charge, –2. So
$$Mg^{+2} \diagdown O^{-2}$$
becomes
$$Mg_2 O_2$$
But we are supposed to have the lowest ratio and we don't. We can divide by 2, which is the common factor. Dividing (or multiplying) all of the members of a ratio by the same number does not change the ratio.

Dividing $Mg_2 O_2$ by 2 gives us the compound named magnesium oxide.
$$MgO$$

All of this is fine, but what about carbon monoxide (CO)? We can predict the oxidation numbers for carbon to be +4 and –4; oxygen would be either a +6 or a –2. There is no combination of charges (+ and –) that will give us a net charge of zero using only one of each.

There is a possible explanation. Remember that helium has two electrons in the first orbit, which is the outside orbit. What would happen if there were two electrons in the outside orbit, but it was not the first orbit? That ion would be an imitator of helium, but a very poor imitator. The effect would be that a compound could be produced and would not be expected to be stable.

Back to carbon monoxide (CO)—the element written on the right in a compound in inorganic chemistry is negative; the element written on the left is positive. If we take oxygen to be negative, the only negative oxidation number predictable is –2. That means that the carbon will have to be +2 to give us a net charge of zero for the compound, CO.

How did carbon become +2? Let's look at the atomic structure of carbon. From thePeriodic Table, we see that the atomic number of carbon is six. Thus, there are six protons and six electrons.

$$\begin{array}{ccc} & \backslash & \backslash \\ 6P & 2 & 4 \\ & / & / \end{array}$$

A carbon atom may be written as

$$\begin{array}{ccc} & \backslash & \backslash \\ 6P & 2 & 2 \,\&\, 2 \\ & / & / \end{array}$$

Suppose only two of the electrons are lost from the outside orbit. Two electrons will be left in the outside orbit and the result will be an imitator of helium.

$$\begin{array}{ccc} & \backslash & \backslash \\ 6P & 2 & 2\,\textcircled{2}\!\longrightarrow \\ & / & / \end{array}$$

The electron loss leaves us with

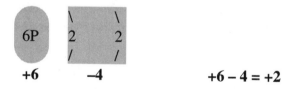

$$\begin{array}{ccc} +6 & -4 & +6-4 = +2 \end{array}$$

True, the imitation is not perfect—the two electrons in the outside orbit are in the second orbit. Remember that helium's outside orbit is the first orbit. We can expect carbon to lose two more electrons; it does. This means that carbon monoxide is not a very stable compound; it will burn in air (oxygen) to produce carbon dioxide (carbon with a charge of +4).

All of Group IVA can display a +2 charge. Groups VA, VIA, and VIIA will have charges that are the result of leaving two electrons in the outside (not the first) orbit. So Group VA can be a +3, Group VIA a +4, and Group VIIA a +5. Group VIIIA doesn't enter into compounds and is considered to have a net charge of zero.

Figure 3.3 presents these alternate positive oxidation numbers and those previously discussed.

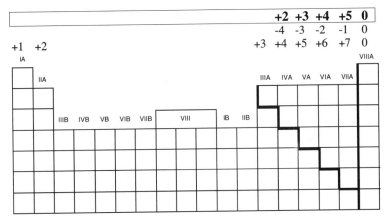

**Figure 3.3
Alternate positive oxidation numbers
for A groups.**

Example 3.5: Using the oxidation numbers of sulfur and oxygen, write the chemical formula for sulfur oxide.

Solution: The oxidation numbers that sulfur (Group VIA) can display are +6, +4, and –2. Oxygen (Group VIA) can be +6, +4, and –2 in a compound.

Generally speaking, the Table tells us which of these is going to be the positive ion and which will be the negative. Elements to the left in a period tend to be more positive; elements lower in a group tend to be more positive.

Sulfur is below oxygen and will play the positive role in the compound. This forces oxygen to be negative since we must have positives and negatives to reach the net charge of zero. So we know that sulfur does not have a –2 charge and that oxygen will not be +4 or +6.

Let's write the symbol of the elements with all the possible charges that can be used.

$$S^{+4 \ \& \ +6} \quad O^{-2}$$

Now we know that there will be two compounds of sulfur and oxygen possible from the charges on the sulfur.

$$S^{+4} O^{-2} \qquad S^{+6} O^{-2}$$

Crossing the charges will give us the two formulas.

$$S_2O_4 \qquad S_2O_6$$

We remove the common factors by dividing S_2O_4 by 2 and S_2O_6 by 2.

$$SO_2 \qquad SO_3$$

SO_2 is named sulfur dioxide and SO_3 is sulfur trioxide.

NOTE: Words like "predict," "tend," and "generally" are sprinkled throughout this chapter and the sciences. Also, you might have noticed that "always" and "never" are not used in this book.

Scientists talk about and use the laws of the sciences. The use of the word "laws" implies that there are no exceptions. In reality, there are exceptions to just about everything except the proven and thoroughly tested laws of science.

*To cover yourself when discussing topics that have exceptions, it
is a good idea to use vocabulary that leaves room for the possibility
of exceptions.*

POLYATOMIC IONS

The major compound found in limestone, chalk, seashells, and coral is $CaCO_3$, calcium carbonate. The formula $CaCO_3$ is made up of three elements, not the two elements per compound we have been discussing.

If we look at the individual elements found in calcium carbonate, we can determine how the elements are behaving.

Calcium (Group IIA) will display only a +2 charge in a compound.

Oxygen (Group VIA) will be +6, +4, or –2 in a compound. (However, the vast majority of the time, oxygen is a –2 in a compound.)

Carbon (Group IVA) will be a +4, +2, or –4 in a compound.

We will have to figure out the charge of carbon in calcium carbonate. We can do that since we know that the net charge (overall charge) on a compound is zero. All we have to do is a bit of addition and subtraction.

$$CaCO_3$$

$$Ca + C + O_3 = 0$$

$$+2 \quad ? \quad -2 \times 3 = 0$$

$$+2 \quad ? \quad -6 = 0$$

The only number that will fit for "?" is +4, the charge of the carbon in this compound. This charge, +4, is a valid charge that we can predict from the Table.

If we look at $CaCO_3$ as a calcium ion hooked up with something else, we will be investigating the CO_3 part of $CaCO_3$.

Since calcium is +2, then CO_3 must be –2 for the net charge of the compound to be zero.

$$Ca^{+2} + CO_3^{-2} = \text{zero net charge}$$

CO_3^{-2} is a **POLYATOMIC ION**, an ion composed of more than one element. Each of the polyatomic ions has a specific name—carbonate, sulfate, phosphate, and others. Each of the polyatomic ions has a specific charge. A partial list is shown in Table 3.1.

NH_4^{+1}	Ammonium	OH^{-1}	Hydroxide
MnO_4^{-1}	Permanganate	ClO_3^{-1}	Chlorate
NO_3^{-1}	Nitrate	SO_4^{-2}	Sulfate
NO_2^{-1}	Nitrite	SO_3^{-2}	Sulfite
$C_2H_3O_2^{-1}$	Acetate	CO_3^{-2}	Carbonate
CN^{-1}	Cyanide	CrO_4^{-2}	Chromate
HCO_3^{-1}	Bicarbonate	PO_4^{-3}	Phosphate

Table 3.1
Selected polyatomic ions.

Writing chemical compounds with the polyatomic ions is done in the same way as we have learned to write compounds with two elements (**BINARY COMPOUNDS**). All we have to do is cross the charges and then make certain we divide through by any common factor.

Example 3.6: What is the formula sodium cyanide?

Solution: The sodium ion is Na^{+1}; the cyanide ion is CN^{-1}.
$$Na^{+1}CN^{-1}$$
Since we do not write ones, the formula is NaCN.

Example 3.7: Write the formula for potassium sulfate.

Solution: Potassium is +1; sulfate is –2. Therefore,
$$K^{+1}SO_4^{-2}$$
The formula of potassium sulfate is K_2SO_4.

Example 3.8: Write the formula for magnesium cyanide.

Solution: This solution is a bit different because there is a problem with writing the formula for the compound. Let's start with the ions and then cross charges.
$$Mg^{+2}CN^{-1}$$
The formula for magnesium cyanide is not $MgCN_2$. The 2 after the CN only refers to the nitrogen and not to the carbon. For the formula to be correct, we must have two cyanides. CN_2 just doesn't do the job, but $(CN)_2$ will. The parentheses around the CN mean that the contents will be multiplied by the 2 and not just the nitrogen.
The formula for magnesium cyanide is $Mg(CN)_2$.

QUESTIONS AND PROBLEMS
1. What is a stable electron configuration?

2. From the standpoint of electron gain and loss, what is the major difference between metals and nonmetals?

3. What is the atomic structure of potassium when it has reached a stable electron configuration?

4. Write the stable electron configuration of aluminum.

5. How many electrons will be in the outside orbit of sulfur and which orbit will that be when sulfur has achieved a stable electron configuration? Assume that the least amount of energy will be expended.

6. Carbon and silicon have a very similar electron gain and loss pattern. Explain the pattern predicted to achieve a stable electron configuration.

7. How many electrons will there be in the outside orbit of radium (at. no. 88) when

it has reached a stable electron configuration and which orbit will that be?

8. What is the number of the electrons in the outside orbit of the following atoms? To which group do the atoms belong? How many electrons will be gained (G) or lost (L) to reach a stable electron configuration? What will be the valence of the ion produced? Fill in the table.

Atom	Group	e⁻ in Out-side Orbit	e⁻ Moved	G/L	Oxidation Number
Na					
Rb					
Mg					
Ca					
Sr					
B					
Ga					
Fr					

9. What are all the predictable valences of sulfur when using the Periodic Table of the Elements?

10. Predict the charges displayed by iodine.

11. Predict the oxidation numbers displayed by silicon in a compound.

12. Write the formulas of the ions named in the following table.

Ion Name	Formula$^{\text{Oxidation Number}}$
Iodide	
Sulfide	
Sulfate	

13. Fill in the missing cells in the following table of ions.

Ion Name	Formula$^{\text{Oxidation Number}}$
O^{-2}	
OH^{-1}	
Potassium	
Phosphate	
NO_3^{-1}	
N^{-3}	
Carbide	
Carbonate	
Bicarbonate	
NH_4^{+1}	
Cyanide	

14. What is the formula for potassium iodide?

15. What is the formula for sodium sulfate?

16. Find the charge on the sulfur in sodium sulfate.

17. Find the charge on the sulfur in sodium sulfite, Na_2SO_3.

18. Fill in the missing cells in the following table of compounds:

Compound Name	Formula
Sodium cyanide	
Sodium carbonate	
Magnesium cyanide	
Magnesium carbonate	
Magnesium phosphate	
Aluminium phosphate	
Aluminium carbonate	
Aluminium cyanide	
Potassium sulfate	
Beryllium nitrate	
	$Ca(MnO_4)_2$
	$NaHCO_3$
	$B(OH)_3$
	NH_4Cl
	$Al_2(SO_3)_3$
	$LiNO_2$
	$Be(CN)_2$
	$(NH_4)_3PO_4$
	$CaCO_3$
	K_2CrO_4

19. What is the formula for the compound(s) of oxygen and carbon?

20. Write the formulas for the compounds of phosphorus and chlorine, nitrogen and oxygen, silicon and sulfur, and arsenic and nitrate.

21. Calculate the charge on the nitrogen in nitric acid.

22. Write the compound boron nitride.

23. Consider nitrous acid, HNO_2. Find the charge on the nitrogen.

24. Find the charge on the sulfur in sulfurous acid, H_2SO_3.

25. What is the formula for calcium sulfide?

26. What is the compound of sulfur and fluorine, assuming they do react with the highest charge on the positive ion?

27. Write the compound of chlorine and fluorine, assuming they react with the highest charge on the positive ion.

28. Write the formula for copper nitrate with the highest charge on the positive ion.

29. What is the formula for iron oxide with the highest charge on the positive ion?

Write the formula for the named compound:

30. _____ Sodium oxide

31. _____ Potassium phosphide

32. _____ Lithium iodide

33. _____ Calcium carbonate

34. _____ Boron acetate

35. _____ Ammonium sulfide

36. _____ Arsenic III fluoride

37. _____ Beryllium phosphate

38. _____ Sodium bicarbonate

39. _____ Magnesium bisulfate

40. _____ Bismuth chloride

41. _____ Aluminum hydroxide

42. _____ Zinc oxide

43. _____ Calcium acetate

44. _____ Sodium hydroxide

45. _____ Nitric acid

46. _____ Sodium cyanide

47. _____ Potassium biphosphate

48. _____ Ammonium chloride

49. _____ Lithium oxide

50. _____ Aluminum carbonate

51. _____ Boron fluoride

52. _____ Sodium sulfide

53. _____ Radium fluoride

54. _____ Beryllium acetate

55. _____ Magnesium diacidphosphate

56. _____ Lead IV hydroxide

57. _____ Ammonium phosphate

58. _____ Potassium bisulfate

59. _____ Sodium carbide

60. _____ Sulfuric acid

61. _____ Acetic acid

62. _____ Atimony IV chloride

63. _____ Silver sulfide

64. _____ Uranium V sulfide

65. _____ Sulfur trixoxide

NOTE TO THE STUDENT

*The ability to write
chemical compounds is
a super-important skill
that is used throughout
all chemistry.
THEREFORE,
practice, practice, practice!*

THE MOLE

• The Mole, a Counting Word •
• Percent Composition • Empirical Formula •
• Molecular Formula •
• Questions and Problems •

THE MOLE, A COUNTING WORD

If you went to the grocery store to buy a dozen eggs, you would expect to get 12 eggs. A jazz quartet is four performers of jazz. A gross of pencils is 144 pencils (1 gross is 12 dozen).

Each of the words—dozen, quartet, gross—represents a number. In the world of chemistry, the **MOLE** is a number of items:

$$6.022 \times 10^{23} \text{ things}$$

Of course, the mole is a rather large number: 602,000,000,000,000,000,000,000 things. That would be about six hundred pink and purple zillion things, which probably has absolutely no meaning. Look at it this way—if you were to count dollar bills at the rate of one each second with no breaks, it would take approximately 20,000,000,000,000,000 years to count them all. It does not appear that the job would get done.

However, the word *mole* is just a word with a number value. True, it is a large numerical value, but it is still just a number. The chemist talks about one mole of atoms in the same manner that you would discuss a dozen eggs; *mole* and *dozen* are words that refer to a number. In other words, be impressed with the mole number, but remember that it is just another counting word and use it that way.

One very important use of mole is related to the information contained on the Periodic Table of the Elements (Table). If we look at the atomic weight of an element, we can express that weight in terms of atomic mass units (amu). The problem with amu's is that the balances in the laboratory do not weigh in amu's. Laboratory balances are calibrated to weigh in grams.

The atomic weight units may be expressed in grams rather than amu's. When we do express atomic weight (or molecular weight) in grams, we are saying that we have one mole of that substance. This means that 22.98977 grams of sodium and 12.011 grams of carbon would contain the same number of atoms, 6.022×10^{23} atoms. How can that be? After all, they do not weigh the same.

Suppose you have a dozen apples and a dozen grapes. Will they weigh the same total weight? Not unless you have really large grapes and/or some very small apples! The same concept applies to atoms (and molecules). A sodium atom has a different number of parts than does a carbon atom. The sodium atom is larger and, logically, each atom of sodium is heavier than each atom of carbon.

Amedeo Avogadro in the early 1800s suggested that equal volumes of gases at the same temperature and pressure contain equal numbers of molecules. This led to the realization that the **GRAM ATOMIC WEIGHT**, the atomic weight in grams, contains one mole of atoms (the same is true for molecules and molecular weights). 6.022×10^{23} things are called one **AVOGADRO'S NUMBER** of things or one **MOLE** of things.

Example 4.1: What is the weight of one mole of sulfur?

Solution: The Periodic Table tells us that the atomic weight of sulfur (S, atomic number 16) is 32.066 amu. We may express that weight in grams, 32.066 grams, which is the weight of one mole of sulfur.

Example 4.2: How much of each of the components of iron II sulfide, FeS, will be required to produce one mole of FeS and how much FeS will we have?

Solution: The atomic weights of iron and sulfur will tell us how much of each to weigh. Iron has the atomic weight of 55.847 $^g/_{mole\ Fe}$ and sulfur's atomic weight is 32.066 $^g/_{mole\ S}$. Since we only need one mole of each for the compound's formula to satisfy the problem,

$$55.847\ ^g/_{mole\ Fe} + 32.066\ ^g/_{mole\ S} = 87.913\ ^g/_{mole\ FeS}$$
$$\text{grams Fe} \qquad \text{grams S} \qquad \text{grams FeS}$$

Example 4.3: If we wanted to produce one mole of sulfuric acid, H_2SO_4, from the elements, what weight of each element would we need and how much sulfuric acid would be produced?

Solution: Sulfuric acid's formula calls for 2 moles of hydrogen, 1 mole of sulfur, and 4 moles of oxygen. We could express the information this way:

	$2 \times H$	$1 \times S$	$4 \times O$
Moles	2×1.0079	1×32.066	4×15.9994
Weight	2.0158 g	32.066 g	63.9976 g

The total weight calculation for H_2SO_4 is
$$2.0158\ \text{g H} + 32.066\ \text{g S} + 63.9976\ \text{g O} = 98.0794\ ^g/_{mole}$$

Example 4.3 tells us that the weight of one mole of H_2SO_4 is 98.0794 g/mole. When we calculate the weight of one mole of a compound, we are calculating the **MOLECULAR WEIGHT**.

Example 4.4: How much sulfuric acid would we have to weigh to obtain 2.5 moles?

Solution: Since we know that H_2SO_4 is 98.0794 $^g/_{mole}$, we can apply dimensional analysis by recognizing that we have to get rid of the moles to get just the weight in grams. We are told to weigh 2.5 moles—if we multiply the weight ($^g/_{mole}$) by moles, we will end up with just the weight:
$$98.0794\ ^g/_{mole} \times 2.5\ \text{moles} = 245.1985\ \text{g}\ H_2SO_4$$

NOTE: The accepted abbreviation for mole is mol.

Sometimes we are asked to calculate the weight of a sample on the basis of the number of moles. You will find that this is common when the mole concept is applied to balanced chemical equations a bit later in this manual.

Example 4.5: Calculate the number of grams found in 20.5 mol of argon.

Solution: The atomic weight of argon (Ar, at. no. 18) is 39.948. We may express

We are required to express the answer in grams. We set up the answer: on the right the equal sign, and the given information on the left.

$$\frac{g\ Ar}{mol\ Ar} = \quad g\ Ar$$

We need to eliminate the moles Ar and do so by multiplying by the moles argon given in the problem.

$$\cancel{mol\ Ar} \times \frac{g\ Ar}{\cancel{mol\ Ar}} = \quad g\ Ar$$

Notice that the units, moles Ar, cancel. This cancellation gives us the answer we need, grams Ar. All we need to do now is place the numbers and calculate.

$$20.5\ \cancel{mol\ Ar} \times \frac{39.948\ g\ Ar}{\cancel{mol\ Ar}} = 818.934\ g\ Ar = 819\ g\ Ar$$

PERCENT COMPOSITION

Imagine this challenge:

> *You are working as a purchasing agent for a company involved in manufacturing. There is a procedure that requires nitrogen in a process. You find out from the chemical engineers that two acids, HNO_3 and HCN, will provide the nitrogen required. The costs of HNO_3 and HCN turn out to be the same per mole. Which of these should you purchase?*

Saving the company as much money as possible will increase the company's profits. Making the best choice of raw materials will certainly make you look good to the people that grant salary raises.

The decision could be made by calculating which of the acids contains the most nitrogen. One way of doing the calculation would be on the basis of the percent nitrogen in the two acids: Ssimply choose the acid with the highest percentage of nitrogen.

Percent composition is calculated in the same manner as any percent; the part we are interested in is divided by the whole and then multiplied by 100.

Example 4.6: Which is a better purchase on the basis of nitrogen, HNO_3 or HCN?

Solution: For the nitric acid, the calculations are as follows:

Nitric acid: HNO_3

	$1 \times H$	$1 \times N$	$3 \times O$
Moles	1×1.0079	1×14.00674	3×15.9994
Weight	1.0079 g	14.00674 g	47.9982 g

We add the total weights of each element to calculate the molecular weight of nitric acid to be 63.01284 $^g/_{mol.}$

To calculate the percent nitrogen in nitric acid, we divide the nitrogen by the complete molecular weight and then multiply by 100:

$$\frac{14.00674\ g\ N}{63.01284\ g\ HNO_3} \times 100 = 22.23\%\ nitrogen\ from\ HNO_3$$

The same the procedure is used to determine the percent nitrogen in HCN so that we may compare the percent composition nitrogen of the two acids:

Hydrocyanic acid: HCN

	1 × H	**1 × C**	**1 × N**
Moles	1 × 1.0079	1 × 12.011	1 × 14.00674
Weight	1.0079 g	12.011 g	14.00674 g

$$\frac{14.00674 \text{ g N}}{27.02564 \text{ g HCN}} \times 100 = 51.83\% \text{ nitrogen from HCN}$$

On the basis of our calculations, HCN is a much better buy than HNO_3. As a matter of fact, your recommendation of HCN would get your company a very good deal for nitrogen since there is over 2.25 times as much nitrogen in HCN as in HNO_3.

Another type of problem presents the percent composition of a compound by element masses and asks us to determine the formula of that compound.

There is a little trap in this type of problem: We cannot determine the molecular formula because we are not given sufficient information. The best that we can do is to express the simplest formula of the compound. The techniques necessary to determine the molecular formula will be covered later in this chapter in the section titled "Molecular Formula."

Example 4.7: Analysis of a compound tells us that the composition is 88.8% Cu and 11.1% O. The percents are expressed in percent of total mass. What is the simplest formula for this compound of copper and oxygen?

Solution: Working this problem depends on the real meaning of percent. Percent is a word that means "parts per hundred" or, if you prefer, "parts out of a hundred."

If we had a 100-g sample of this compound, we would have 88.8 g copper and 11.1 g oxygen. The total weight of the sample from the percentages would 99.9 grams—very close to 100 grams. The total weight may not be exactly 100, but will be very close.

Knowing the weights of the elements present, we can easily calculate the number of moles of each present. WHY? The reason for the calculation is that *chemical compound formulas are written as a mole ratio*.

Let's start with copper. 88.8% Cu is 88.8 g Cu, assuming a 100-g sample. We can use dimensional analysis to convert from grams copper to moles copper. Remember that one gram atomic weight is one mole of the element. In the case of copper, that is

63.546 g / mole Cu

OR

$$\frac{63.546 \text{ g Cu}}{1 \text{ mol Cu}}$$

OR

$$\frac{1 \text{ mol Cu}}{63.546 \text{ g Cu}}$$

Since we have 88.8 g Cu and need to get rid of the "g Cu" to get to moles Cu, we can multiply to cancel units using dimensional analysis:

$$88.8 \text{ g Cu} \times \frac{1 \text{ mol Cu}}{63.546 \text{ g Cu}} = 1.3974 \text{ mol Cu} = 1.4 \text{ mol copper}$$

We can certainly do the same thing for the oxygen:

$$11.1 \text{ g O} \times \frac{1 \text{ mol O}}{15.9994 \text{ g O}} = 0.6938 \text{ mol O} = 0.69 \text{ mol oxygen}$$

We now have a mole ratio of copper:oxygen. The compound would be $Cu_{1.4}O_{0.69}$. This formula is not acceptable; all chemical formulas must be written using whole numbers.

This is not a problem. If we are to divide the number of moles of each element by the smallest number of moles, 0.6938, we may well be able to produce a whole number ratio:

For copper: 1.3974/0.6938 = 2.014
For oxygen: 0.6937/0.6938 = 1.0

The value for copper is so close to 2 that we may round off to 2. Therefore, the ratio of 2 Cu:1 O tells us that the simplest formula of the compound is Cu_2O.

EMPIRICAL FORMULA

Example 4.7 has us writing the simplest formula for a compound. The simplest formula is called the **EMPIRICAL FORMULA**.

Let's take a look at the formula for glucose, $C_6H_{12}O_6$. If we divide through this molecular formula by six, we end up with $C_1H_2O_1$. Of course, we don't write ones unless we absolutely have to; ones are not required in chemical formulas. Therefore, the basic formula for glucose, when expressed with the common factor of 2 removed, is CH_2O.

When a chemical formula is expressed with all common factors removed, that formula is the simplest expression of the formula and is the empirical formula.

Sometimes we are presented with an unknown compound and are asked to determine the formula for the compound. The first step is to weigh the sample carefully. The second step is to decompose the compound or react the compound with oxygen (burn it). We weigh the breakdown products or compounds produced and determine the empirical formula of the original compound.

We cannot determine the molecular formula unless we know the molecular weight, as is shown in the next section.

Example 4.8: A 0.5000-g sample of a compound was found to contain 0.3619 g iron and 0.1381 g oxygen. Calculate the empirical formula.

Solution: We cannot write the compound formula based on grams composition; compound formulas are written as a mole ratio of the elements. We already have the formula for this compound:

$$Fe_xO_y$$

All we have to do is determine the mole ratio, *x:y*.

We can convert the masses of the elements to moles. Then we can write the formula. Since iron is mentioned first, let's start with iron. We need to convert 0.3619 grams Fe to moles Fe. This can be done by getting rid of the grams Fe and going to moles Fe. We use dimensional analysis and the Periodic Table place values.

$$0.3619 \text{ g Fe} \times \frac{1 \text{ mol Fe}}{55.874 \text{ g Fe}} = 6.477 \times 10^{-3} \text{ mol Fe}$$

We now use the same technique to determine the information for the number of moles oxygen.

$$0.1381 \text{ g O} \times \frac{1 \text{ mol O}}{15.9994 \text{ g O}} = 8.632 \times 10^{-3} \text{ mol O}$$

Our mole ratio is 6.477×10^{-3} iron:8.632×10^{-3} oxygen (or as 6.477:8.632 if we remove the common factor of 10^{-3} by dividing through by 10^{-3}). We can write a chemical formula at this point.

$$Fe_{6.477}O_{8.632}$$

Of course, this formula won't work because chemical formulas are a whole number relationship. So we need to do a bit more work.

If we divide both of the members of the ratio by the smallest number, we will be on the road to a whole number relationship. The reason we are able to do this is that the ratio is not changed if we divide or multiply all members of the ratio by the same number. *This concept will not work if we add or subtract, only if we multiply or divide.*

$$6.477 \times 10^{-3} \text{ mole Fe}/6.477 \times 10^{-3} = 1 \text{ mole Fe}$$
$$8.632 \times 10^{-3} \text{ mole O}/6.477 \times 10^{-3} = 1.33 \text{ mole O}$$

We can write $Fe_1O_{1.33}$, but we still need a whole number ratio. We can multiply by 3 because of the .33 in 3.33 in order to achieve a whole number ratio of 3:4. Therefore, the simplest formula for this compound is Fe_3O_4.

Summary Solution:

For the iron,

$$0.3619 \text{ g Fe} \times \frac{1 \text{ mol Fe}}{55.874 \text{ g Fe}} = 6.477 \times 10^{-3} \text{ mol Fe}$$

For the oxygen,

$$0.1381 \text{ g O} \times \frac{1 \text{ mol O}}{15.9994 \text{ g O}} = 8.632 \times 10^{-3} \text{ mol O}$$

The mole ratio is 6.477×10^{-3} iron:8.632×10^{-3} oxygen or 6.477:8.632.

$$Fe_{6.477}O_{8.632}$$

To achieve a whole number ratio divide by the smallest value.

$$6.477 \times 10^{-3} \text{ mole Fe}/6.477 \times 10^{-3} = 1 \text{ mole Fe}$$
$$8.632 \times 10^{-3} \text{ mole O}/6.477 \times 10^{-3} = 1.33 \text{ mole O}$$

To produce a whole number ratio, multiply by 3.

$$3(1 \text{ mole Fe}:1.33 \text{ mole O})$$
$$3 \text{ moles Fe}:4 \text{ moles O}$$

The simplest formula for this compound is Fe_3O_4.

MOLECULAR FORMULA

Suppose that we are given an unknown sample of a substance for analysis and are asked to determine the molecular formula. The techniques we need to use will be the same as in the previous section. The only difference is that we will have to determine the molecular weight. There is some very elaborate laboratory equipment that can do this for us.

The analysis is aimed toward determining the components and the amounts of those components in the sample. Armed with this information, we can calculate the empirical formula. Then all we have to do is determine how many empirical formula weights are required to equal the molecular weight.

Example 4.9: In the section "Empirical Formula" we discussed glucose, $C_6H_{12}O_6$, as having the empirical formula of CH_2O. How many empirical formulas are required to have the molecular formula?

Solution: We know that the molecular formula for glucose is $C_6H_{12}O_6$. We also know that the empirical formula for glucose is CH_2O. The common factor of 6 may be removed from the molecular formula to express the empirical formula, CH_2O. Therefore, there are six empirical formulas required for the molecular formula of glucose.

Alternate Solution: The molecular formula weight (MFW) of $C_6H_{12}O_6$ is 180.1572 g $C_6H_{12}O_6$. The empirical formula weight (EFW) of CH_2O is 30.0262 g CH_2O. If we divide the molecular formula weight by the empirical formula weight, we will know how many empirical formulas are required.

$$\frac{180.1572 \text{ }^g\!/_{MFW} \text{ } C_6H_{12}O_6}{30.0262 \text{ }^g\!/_{EFW} \text{ } CH_2O}$$

The calculation tells us that we need six EFs for each MF.

Example 4.10: A compound was found to have a molecular weight of 114.23 g. A 1.50-g sample of the compound was tested to determine its elements. The analysis provided us with 12.62 g carbon and 2.38 g hydrogen. What is the molecular formula for this compound?

Solution: We are required to write the formula for C_xH_y. First we calculate the empirical formula (EF), and then we determine how many EFs we still need to write the molecular formula (MF).

For the carbon,

$$12.62 \text{ g C} \times \frac{1 \text{ mol C}}{12.011 \text{ g C}} = 1.051 \text{ mol C}$$

For the hydrogen,

$$2.38 \text{ g H} \times \frac{1 \text{ mol H}}{1.0079 \text{ g H}} = 2.363 \text{ mol H}$$

The mole ratio generated is

1.051 moles C : 2.363 moles H

The calculation gives us the compound $C_{1.051}H_{2.363}$. However, we do need a whole number ratio. If we divide through the ratio by the smallest number, 1.051, we take a step toward a whole number ratio.

$$1 \text{ mole } C : 2.248 \text{ moles } H$$

Since 2.248 is very close to 2.25, we can multiply the mole ratio by 4 to finally reach a whole number ratio. The simplest formula is C_4H_9.

The empirical formula weight is 57.1151 and the problem statement tells us that the molecular weight is 114.23. If we divide the molecular weight by the empirical formula weight, we will know how many EFs we need per MF.

$$114.23 \text{ g}/57.1151 \text{ g} = 2 \text{ EFs per MF}$$

$(C_4H_9)_2$ becomes C_8H_{18}, the molecular formula.

One method used for the analysis of organic compounds (compounds containing carbon) is to burn the compound and collect the products for analysis. When this technique is used, it sometimes appears as though we do not have sufficient information to determine the composition by elements, especially for oxygen.

The products of the burning are compounds containing oxygen (CO_2, H_2O, etc.). We can determine the amount of the elements (carbon, hydrogen, etc.) from the original compound by applying the percent composition calculation to isolate each element by mass from the compound. If the compound contains oxygen, as does glucose, we have to work backward, as shown in Example 4.11.

Example 4.11: A 0.5562-g sample of a compound containing carbon, hydrogen, and oxygen is burned; 0.8291 g CO_2 and 0.2546 g H_2O are collected. The molecular weight is found to be 118.089 g. What is the molecular formula for this compound?

Solution: The compound we are asked to write is $C_xH_yO_z$. We are provided with enough information to calculate the carbon and hydrogen if we use the percent composition of carbon in CO_2 and hydrogen in H_2O. However, we are given no direct information about the oxygen. We can deduce the mass of oxygen in the original sample by subtracting the weight of the carbon and hydrogen from the weight of the sample. First, we must calculate the mass of carbon and hydrogen from the products of the burning.

CO_2 (0.8291 g CO_2) is collected. The percentage of carbon in CO_2 is calculated by means of

$$\frac{C}{CO_2} \times 100 = \% \ C$$

$$\frac{12.011 \text{ g C}}{44.0098 \text{ g } CO_2} \times 100 = 17.292\% \ C$$

Then, we take 27.292% of the mass of the CO_2 collected to calculate the carbon.

$$0.8291 \text{ g } CO_2 \times \frac{0.27292 \text{ g C}}{1 \text{ g } CO_2} = 0.2263 \text{ g C}$$

Notice that the above process is a two-step process. The problem can be worked in one step; we could work with the fraction of carbon found in carbon dioxide and then multiply.

$$0.8291 \text{ g } CO_2 \times \frac{12.011 \text{ g C}}{44.0089 \text{ g } CO_2} = 0.2263 \text{ g C}$$

Of course, we have to do the same calculations to isolate the mass of the hydrogen collected in the water.

$$0.2546 \ \text{g } H_2O \times \frac{2.0158 \ \text{g H}}{18.0182 \ \text{g } H_2O} = 0.285 \ \text{g H}$$

Subtracting the masses of carbon and hydrogen from the sample provides us with the amount of oxygen in the sample.

$$0.5562 \text{ g sample} - 0.2263 \text{ g C} - 0.0285 \text{ g H} = 0.3014 \text{ g O}$$

Working from the masses of the components of the compound, we derive the moles of each so we may write the compound formula.

$$0.2263 \ \text{g C} \times \frac{1 \ \text{mol C}}{12.011 \ \text{g C}} = 0.01884 \ \text{mol C}$$

$$0.0285 \ \text{g H} \times \frac{1 \ \text{mol H}}{1.0079 \ \text{g H}} = 0.02828 \ \text{mol H}$$

$$0.3014 \ \text{g O} \times \frac{1 \ \text{mol O}}{15.9994 \ \text{g O}} = 0.01884 \ \text{mol O}$$

The formula cannot be written quite yet as we need a whole number mole relationship. We can divide all of the elements of the ratio by the smallest number of moles without changing the ratio. We get 1 C:1.5 H:1 O. If we multiply all of the elements of the mole ratio by a number, we do not change the mole ratio. Notice that there are 1.5 mol hydrogen. Then we multiply the mole ratio (1 C:1.5 H:1 O) by 2. The multiplication gives us with the mole ratio of 2 C:3 H:2 O; this is the ratio of the empirical formula: $C_2H_3O_2$.

We were told by the problem that the molecular weight of the compound is 118.089. $C_2H_3O_2$ weighs only 59.0445 g. If we divide the empirical formula weight into the molecular formula weight, we will know how many empirical formulas we must have.

$$118.089 \ ^g\!/_{MF} \ / \ 59.0445 \ ^g\!/_{EF} \ = \ 2 \ ^{EF's}\!/_{MF}$$

We now have the formula for the compound.

$$(C_2H_3O_2)_2 \ = \ C_4H_6O_4$$

Solution Summary: The answer to the problem is $C_xH_yO_z$.

The C and H masses come from the CO_2 and H_2O.

0.8291 g CO_2 were collected. The mass of carbon in CO_2 is

$$0.8291 \ \text{g } CO_2 \times \frac{12.011 \ \text{g C}}{44.0089 \ \text{g } CO_2} = 0.2263 \ \text{g C}$$

The hydrogen calculation is

$$0.2546 \ \text{g } H_2O \times \frac{2.0158 \ \text{g H}}{18.0182 \ \text{g } H_2O} = 0.285 \ \text{g H}$$

To calculate the oxygen in the sample,

$$0.5562 \text{ g sample} - 0.2263 \text{ g C} - 0.0285 \text{ g H} = 0.3014 \text{ g O}$$

To obtain a mole ratio,

$$0.2263 \ \text{g C} \times \frac{1 \ \text{mol C}}{12.011 \ \text{g C}} = 0.01884 \ \text{mol C}$$

$$0.0285 \ \cancel{g\,H} \times \frac{1 \ \text{mol H}}{1.0079 \ \cancel{g\,H}} = 0.02828 \ \text{mol H}$$

$$0.3014 \ \cancel{g\,O} \times \frac{1 \ \text{mol O}}{15.9994 \ \cancel{g\,O}} = 0.01884 \ \text{mol O}$$

Divide by the smallest number in an attempt to reach a whole number ratio.

1 C:1.5 H:1 O

Multiply by 2 due to the 1.5 to get the empirical formula, $C_2H_3O_2$.

The MW of the compound was 118.089. $C_2H_3O_2$ weighs 59.0445 g. Divide the EF weight into the MW to know how many EFs we must have.

$$118.089 \ ^g\!/_{MF} \ / \ 59.0445 \ ^g\!/_{EF} \ = \ 2 \ ^{EF's}\!/_{MF}$$

The formula for the compound is $(C_2H_3O_2)_2 \ = \ C_4H_6O_4$.

QUESTIONS AND PROBLEMS

1.   Calculate the weight of one mole of uranium.

2.   What is the weight of one mole of iron?

3.   Calculate the mass of 18 moles of sodium.

4.   Express the weight of 0.255 moles of copper.

5.   Calculate the molecular weight of calcium chloride.

6.   What is the molecular weight of lead IV bromide?

7.   Calculate the weight of 3 moles of sulfuric acid.

8.   What is the weight of 0.625 moles of nitric acid?

9.   Calculate the weights of lithium and iodine required to produce one mole of lithium iodide with no excess of either reagent.

10.   How many grams of each element are required to produce one mole of magnesium sulfide?

11.   What mass of each of the components is required to produce one mole of carbonic acid?

12.   How much of each of the elements do we need to react to produce one mole of nitric acid?

13.   Calculate the % Ag in AgCl.

14.   Calculate the percent carbon in the compound sodium bicarbonate.

15.   What is the percent sulfur in copper II sulfate?

16. Which contains more sulfur, sulfurous acid or sulfuric acid? Express as a ratio.

17. Consider copper II nitrate and copper II carbonate. Which has a higher percentage of copper?

18. We have a choice of purchasing $MgCO_3$ or $MgCl_2$ at the same price per kilogram. On the basis of the magnesium content, which is a better buy?

19. A compound is found to be 42.88% carbon and 57.12% oxygen. Write the simplest formula for the compound.

20. A compound is decomposed and analyzed, the compound is 63.54% iron and 36.46% sulfur. What is the empirical formula?

21. A sample of a compound is decomposed. The results of the decomposition are 2.809 grams of silicon and 3.1999 grams of oxygen. What is the empirical formula for this compound?

22. A 6.55-gram sample of a hydrocarbon, C_xH_y, is burned completely. The products are 19.61 g CO_2 and 10.71 g H_2O. Calculate the empirical formula.

23. 0.750 grams of a hydrocarbon are burned in an excess of oxygen. 2.3532 g CO_2 and 0.9633 g H_2O are produced. Write the empirical formula.

24. A 2.5500-gram sample of a compound known to be a carbohydrate, $C_xH_yO_z$, is burned in an excess of oxygen. 3.6162 g CO_2 and 2.2204 g H_2O are collected. What is the simplest formula for the carbohydrate?

25. What is the empirical formula of a carbon-hydrogen-oxygen compound if 0.4305 grams are burned to produce 0.8803 g CO_2 and 0.2703 g H_2O?

26. A compound of calcium and carbon, 15.36 grams, reacts with oxygen. 18.691 g CaO and 7.334 g CO_2 are collected. What is the empirical formula?

27. Fifteen grams of a compound known to be a hydrocarbon with the molecular weight of 42.084 are burned in an excess of oxygen; 47.06 g CO_2 and 19.27 g H_2O are collected. Find the molecular formula.

28. Two grams of a liquid hydrocarbon with a molecular weight of 114.23 are burned completely. The compounds collected are CO_2 (6.1644 g) and H_2O (2.8388 g). What is the molecular formula?

29. A carbohydrate, $C_xH_yO_z$, is burned. A 2.60-g sample produces 3.81085 g CO_2 and 1.5601 g H_2O. Write the molecular formula (MW = 120).

30. Four grams of a carbon-hydrogen-oxygen compound, MW = 46, are burned; 4.70 grams of water and 7.65 grams of carbon dioxide are produced. Determine

the molecular formula.

31. A hydrocarbon has the molecular weight of 58.123. A 45.00-g sample is reacted with an excess of gaseous chlorine. The products of the reaction are 476.368 g CCl_4 and 282.288 g HCl. What is the molecular formula of the hydrocarbon?

32. A compound is composed of carbon, hydrogen, and lead. The compound has the molecular weight of ±969. The products from the reaction of a 225-g sample and an excess of chlorine are 856.029 grams CCl_4, 507.265 grams HCl, and 242.784 grams $PbCl_4$. Determine the molecular formula of the compound.

33. Find the empirical formula of a copper-sulfur compound on the basis of the following data: A 10.00-gram sample combined with oxygen produces 8.3196 g CuO and 8.3739 g SO_3.

34. A 33-g sample of a compound known to be composed of nitrogen and oxygen reacts with fluorine. 78.2 g of NF_5 and 186.5 g of OF_6 are collected. The molecular weight of the nitrogen-oxygen compound is 92 g/mol. What is the molecular formula?

35. A 6.00-g sample of a uranium-sulfur compound is reacted with fluorine. The compounds collected are UF_5, 6.2796 grams, and SF_6, 6.8853 grams. What is the empirical formula of the uranium-sulfur compound? Is the empirical formula also likely to be the molecular formula?

NOTE TO THE STUDENT

*A thorough understanding of, and ability to use
the mole is very important throughout Chemistry.
THEREFORE,
Practice, practice, practice.*

• What Is Balancing and Why? • Combination • Decomposition •
• Single Replacement • Double Replacement •
• Oxidation-Reduction (Redox) • Balancing Redox Reactions •
• Questions and Problems •

WHAT IS BALANCING AND WHY?

Balancing a chemical equation is an absolute requirement of the laws of chemistry and physics. According to the law of conservation of matter, matter can neither be created nor may it be destroyed by normal chemical means. Balancing a chemical equation is a method of accounting for the matter.

Balancing an equation is much like watching children play on a seesaw at the park. To get the seesaw to balance, the weight of children on one side has to equal the weight of children on the other side. Another place where balancing is applied is the scales of justice used as a symbol in the practice of law (Fig. 5.1).

Figure 5.1
Scales of justice.

Suppose we were to look at the equation by which iron reacts with sulfur to produce iron II sulfide:

$$Fe + S \rightarrow FeS$$

We could read this equation to be "one atom of iron reacts with one atom of sulfur to produce one molecule of iron II sulfide." This reading of the equation does account for the conservation of the matter.

Another way of looking at the equation is based on the weights of the materials present. Using the rounded-off weights from the Periodic Table, we can read the equation as "one mole, 56 grams iron, reacts with one mole sulfur, 32 grams sulfur, producing one mole of ironIIsulfide, 88 grams FeS."

$$Fe + S \rightarrow FeS$$
$$\textit{1 mole} + \textit{1 mole} \rightarrow \textit{1 mole}$$
$$56 \text{ g Fe} + 32 \text{ g S} \rightarrow 88 \text{ g FeS}$$

By using any of these methods to look at the equation, we can easily account for all the matter involved. We see that no matter is gained; nor is any lost.

Looks pretty simple . . . it is! Suppose we look at an equation that appears to be more complex. Let's investigate the reaction for the production of sodium sulfide from sodium and sulfur.

$$Na + S \rightarrow Na_2S$$

This equation is *not* balanced. Notice that there is one sodium on the left, but there are two on the right side. The way to balance this equation is to have two sodiums on the left.

$$2Na + S \rightarrow Na_2S$$

We can balance this equation by doubling the sodium. **2Na** is used, not **Na$_2$**. If we double the sodium by writing **Na$_2$**, we have messed around within the formula for sodium; we can't do that. The only way to double the sodium is to double the entire formula.

What is the procedure for balancing an equation? As with most things in the sciences, there is a pattern to balancing equations. The steps can be written as follows:

1. Write all formulas correctly.
2. Start with the left-most atom and balance with the right side of the equation. *Do not make changes within the formula. Changes within a formula will change the ratio of elements, which will either give a different compound or something that does not exist. Multiplying the entire formula is the method to get the **proper** amount of matter required.*
3. Go to the next left-most atom and balance.
4. Continue until all atoms have been balanced.
5. Double check the balancing using the same pattern of Steps 1–4.

Example 5.1: Write and balance the chemical reaction in which calcium reacts with sulfur to produce calcium sulfide.

Solution: Using Step 1, we write the participants in the reaction.

$$Ca + S \rightarrow CaS$$

Following Step 2, we look to see how many moles there are of calcium on both sides. We find that there is one mole calcium on each side of the equation. Calcium is balanced. Step 3 requests that we look at the next atom to be balanced; that is the sulfur. There is one mole of sulfur on both sides. The sulfur is balanced.

A double check of the reaction shows that the calcium is in balance and that the sulfur is in balance.

The equation is balanced.

We read the equation as "one mole of calcium will react with one mole of sulfur to yield (produce) one mole of calcium sulfide."

Example 5.2: What is the balanced equation for the reaction of francium and sulfur to produce francium sulfide?

Solution: The first thing to do is to write the reactants and products correctly.

$$Fr + S \rightarrow Fr_2S$$

Then we balance the left-most atom, francium. There is one francium on the left and there are two on the right. We need to have two franciums on the left and may do so by doubling the francium.

$$2Fr + S \rightarrow Fr_2S$$

*Note: Notice that **franciums** are balanced without even thinking in moles, atoms, or whatever. This will work because we may consider moles and atoms interchangeably as long as we are accounting for the matter involved in the equation.*

Next we need to look at the sulfur. There is one sulfur on the left and one on the right—balanced.

Double checking the equation indicates that there are two franciums on both sides and one sulfur on both sides. This equation is balanced.

In Example 5.2, we could not double the francium on the left by writing **Fr_2**. The reason is that francium is not a **DIATOMIC** element. Some elements are diatomic (two atoms to the molecule) when found free in nature. The diatomic elements are hydrogen (Group IA), nitrogen (Group VA), oxygen (Group VIA), and the Group VIIA **HALOGENS** (fluorine, chlorine, bromine, and iodine). Find these elements on the Table; notice that there is a definite pattern making it easy to remember them.

It is important to notice that the diatomic elements are diatomic, two atoms to the molecule, when they are **FREE**. *Free* means that they are not combined with other elements. For example, hydrogen is diatomic when free—H_2. The formula for water, H_2O, contains two hydrogens because that is what is required to write the compound. The hydrogen (and the oxygen) are not free; the hydrogen is chemically bound to the oxygen in the water molecule. Molecules must have a zero charge.

Example 5.3: Write the balanced equation for the formation of water from the elements.

Solution: We first need to write the equation's reactants and products correctly.

$$H_2 + O_2 \rightarrow H_2O$$

We must write hydrogen and oxygen as diatomic on the left side of the equation. They are free; they are not chemically bound to anything. However, water is written in a two to one ratio in order to get an overall charge on the molecule of zero.

Next we start on the extreme left. Notice that there are two hydrogens on both sides. Hydrogen is balanced as written. Then we look at the oxygen. There are two on the left and one on the right. We need to double the oxygen on the right to balance. We cannot write H_2O_2 because that will change the ratio of elements within the compound. As a matter of fact, H_2O_2 is hydrogen peroxide, not water. We can double the oxygen in water by doubling the water.

$$H_2 + O_2 \rightarrow 2H_2O$$

Since we have balanced each element once, we now need to double check the equation. We start on the left and find that the hydrogen is now out of balance. There are two on the left and four on the right. So we need to double the hydrogen on the left.

$$2H_2 + O_2 \rightarrow 2H_2O$$

The next element is oxygen. There are two on the left and on the right. The oxygen is in balance.

We need to double check. There are four hydrogens on both sides; there are two oxygens on both sides. The equation is balanced.

The balanced equation may be read as "two moles of hydrogen react with one mole of oxygen to yield (produce) two moles of water."

NOTE: Keep in mind that doubling of the water by $\underline{2}H_2O$ means that there are two waters ($2H_2O = H_2O + H_2O$). There are four hydrogens and two oxygens present.

COMBINATION

Combination reactions are exactly as the name implies. Elements react with each other to produce compounds that are a combination of those elements. Examples 5.1, 5.2, and 5.3 are combination reactions.

Example 5.4: One of the reactions involved in rusting is the production of iron III oxide. Write the equation for the formation of iron III oxide from the elements.

Solution: The first step is to write everything involved.

$$Fe + O_2 \rightarrow Fe_2O_3$$

The left-most element is iron. There is one iron on the left, but there are two on the right. We have to double the iron on the left to balance.

$$2Fe + O_2 \rightarrow Fe_2O_3$$

The oxygen is next. There are two on the left and three on the right. The way to balance is with the lowest number containing the common factors of two and three. That number is six. Twelve would work since it contains both two and three, but twelve is not the *lowest* number. Remember that we have to double and triple the entire molecule and cannot change the ratio of elements within the molecule.

$$2Fe + 3O_2 \rightarrow 2Fe_2O_3$$

Double checking shows us that the iron is out of balance now. There are two irons on the left and four on the right. If we double the two irons on the left, we have four to balance with the right.

$$4Fe + 3O_2 \rightarrow 2Fe_2O_3$$

When we double check again, we find that there are four irons on both sides and that there are six oxygens on both sides. The equation is balanced.

We may read this equation as "four moles of iron react with three moles of oxygen to yield two moles of iron III sulfide."

DECOMPOSITION

Decomposition reactions are exactly like combination reactions except that they are the reverse.

In Example 5.3, we produced water from the elements.

$$2H_2 + O_2 \rightarrow 2H_2O$$

Suppose we passed an electrical current through water from a battery (direct current source). The water molecule would decompose to the component elements.

$$2H_2O \rightarrow 2H_2 + O_2$$

This equation reads "two moles of water will decompose to two moles of hydrogen and one mole of oxygen."

Example 5.5: A Downs cell is used to decompose table salt to its elements. What is the balanced chemical reaction?

Solution: First, we write the chemical reaction:

$$NaCl \rightarrow Na + Cl_2$$

Now we need to check the sodium. There is one on both the left and the right.

The chlorine is not in balance. There is one on the left but two on the right. We need to double the sodium chloride to double the chlorine.

$$2NaCl \rightarrow Na + Cl_2$$

A double check shows us that the sodium is now out of balance. If we double the sodium on the right, we will bring sodium into balance.

$$2NaCl \rightarrow 2Na + Cl_2$$

We double check the work and find that the equation is balanced.

We read this equation as "two moles of sodium chloride decompose to release two moles of sodium and one mole of chlorine."

59

Chapter 5
Chemical Reactions

SINGLE REPLACEMENT

In the single replacement reaction, an element within the compound is replaced from an element from outside the compound. The reason that these reactions proceed is that the atom being replaced is not as chemically active as the element entering the compound.

One way of predicting chemical activity is to look at the Periodic Table of the Elements. The general rules that predict chemical activity are as follows:

1. The lower in a group, the greater the chemical activity as a metal. Metals display a posative (+) charge in a compound.
2. The further to the left in a period, the greater the chemical activity as a metal (+).
3. The higher in a Group IVA through Group VIIA, the greater the chemical activity as a nonmetal (–).
4. The further to the right in a period, the greater the chemical activity as a nonmetal (–).

*Please keep in mind that these rules are only for **predictions**. There are exceptions that are not detailed in this chapter.*

Example 5.6: Write the balanced equation for the reaction of magnesium and hydrochloric acid.

Solution: As with the preceding examples, we write the reactants and products.
$$Mg + HCl \rightarrow MgCl_2 + H_2$$
We first check the magnesium and find that the magnesium is balanced.
$$Mg + HCl \rightarrow MgCl_2 + H_2$$
The next element is hydrogen. There is one on the left, and there are two on the right. We should double the hydrogen in the hydrochloric acid. Remember that we can only double the HCl; we cannot just double the hydrogen *within* the HCl.
$$Mg + 2HCl \rightarrow MgCl_2 + H_2$$
The third element, chlorine, is in balance with two chlorines on both sides. We find that the equation is balanced when we double check.

NOTE: Most metals will replace hydrogen in an acid.

IMPORTANT: The products of a metal/acid reaction are a salt and hydrogen.

Example 5.7: Finely divided (powdered) francium metal is mixed with finely divided calcium carbonate.
(a)   Will there be a chemical reaction?
(b)   If there is a predictable chemical reaction, write and balance.

Solution:
(a)   We will predict that a chemical reaction will occur. Francium is to the left of calcium and is below calcium on the Table.

(b)   The products and reactants are
$$Fr + CaCO_3 \rightarrow Fr_2CO_3 + Ca$$
To balance the francium, we proced as follows: Since there is one on the left and there are two on the right side of the equation, we double the francium on the left.
$$2Fr + CaCO_3 \rightarrow Fr_2CO_3 + Ca$$
The next element to balance is the calcium. There is one on the left and on the right; calcium is balanced.

Carbons? There is one carbon on the left and one carbon on the right.

There are three oxygens on both sides.

A double check of the equation tells us that it is balanced.

SHORTCUT: If we noticed that carbon and oxygen behaved in the same ratio, 1:3, on both the left and right of the equation, we could have balanced carbonate against carbonate.

DOUBLE REPLACEMENT (Metathesis)

If single replacement reactions are compared to dancing, then the reaction is like someone cutting in on one of the partners. Double replacement reactions, when considered as dancing, are like the members of two couples switching partners.

Example 5.8: What is the balanced chemical equation for the reaction of hydrochloric acid and potassium hydroxide?

Solution: The basic chemical equation is written as
$$HCl + KOH \rightarrow KCl + HOH$$
Your questions is, "What is HOH and why?" Using HOH gives us the chance to balance on the basis of the **H** from the acid and the **OH** from the base. This is much easier than balancing **H** and **OH** against H_2O. The reason is that the hydrogen is in two places on the left and would be more difficult to balance against H_2O than HOH. Try balancing both ways and you will more than likely prefer HOH to H_2O in these equation types.
$$\underline{H}Cl + KOH \rightarrow KCl + \underline{HOH}$$
There is one hydrogen on the left and right. There is one chloride on the left and one on the right. There is one potassium on each side of the equation. There is one hydroxide on both sides of the equation. The equation is balanced.
$$HCl + KOH \rightarrow KCl + HOH$$
This equation states that "one mole of hydrochloric acid will react with one mole of potassium hydroxide to produce (yield) one mole of the salt, potassium chloride, and one mole of water."

IMPORTANT: The products of an acid/base reaction are a salt and water.

Let's try an equation that looks a great deal more complex but really isn't.

Example 5.9: Write and balance the equation for the reaction between phosphoric acid and magnesium hydroxide.

Solution: The reaction to be balanced is written as

$$H_3PO_4 + Mg(OH)_2 \rightarrow Mg_3(PO_4)_2 + HOH$$

Since hydrogen and hydroxide from the left show up in the water on the right, let's balance hydrogen against hydrogen and hydroxide against hydroxide, as we did in Example 5.8.

$$\underline{H}_3PO_4 + Mg(\mathbf{OH})_2 \rightarrow Mg_3(PO_4)_2 + \underline{H}OH$$

The first check is for the hydrogen; hydroxide will be balanced separately. There are three hydrogens on the left and only one on the right. How about tripling the water on the right?

$$\underline{H}_3PO_4 + Mg(\mathbf{OH})_2 \rightarrow Mg_3(PO_4)_2 + 3\underline{H}OH$$

Notice that the phosphate stays together on the right side. We can balance phosphates and not have to worry about balancing the elements separately—much easier and quicker. We have to double the phosphate on the left.

$$2\underline{H}_3PO_4 + Mg(\mathbf{OH})_2 \rightarrow Mg_3(PO_4)_2 + 3\underline{H}OH$$

Magnesium is next. There is one on the left, and there are three on the right. Let's triple the magnesium hydroxide.

$$2\underline{H}_3PO_4 + 3Mg(\mathbf{OH})_2 \rightarrow Mg_3(PO_4)_2 + 3\underline{H}OH$$

One look at the hydroxide tells us that there are six on the left and three on the right. If we double the amount of hydroxide on the right, we will have the six we need to balance hydroxides.

$$2H_3PO_4 + 3Mg(OH)_2 \rightarrow Mg_3(PO_4)_2 + 6HOH$$

A double check of our work indicates that we have a balanced equation.

The equation reads as follows: "Two moles of phosphoric acid react with three moles of magnesium hydroxide to yield one mole of magnesium phosphate and six moles of water."

Example 5.10: Would there be a reaction if sodium phosphate and barium nitrate solutions were to be mixed? If the reaction does occur, write and balance.

Solution: If the two were to switch partners in the chemical dance, one of the products would be barium phosphate. Barium phosphate is insoluble and will precipitate, leaving sodium nitrate in solution. *The formation of the precipitate is what drives this reaction to proceed.*

Now we write the equation to be balanced.

$$Na_3PO_{4\,(aq)} + Ba(NO_3)_{2\,(aq)} \rightarrow Ba_3(PO_4)_{2\,(s)} + NaNO_{3\,(aq)}$$

*NOTE: **aq**, aqueous, indicates that the compound is in solution. **s**, solid, refers to the precipitate that is not in solution.*

We start with the sodium. There are three on the left and one on the right. We should triple the sodium nitrate to balance the sodium.

$$Na_3PO_{4\,(aq)} + Ba(NO_3)_{2\,(aq)} \rightarrow Ba_3(PO_4)_{2\,(s)} + 3NaNO_{3\,(aq)}$$

Phosphate is next as it stays together on both sides. We need to have two phosphates on the left to balance with the right. We double the sodium phosphate:

$$2Na_3PO_{4\,(aq)} + Ba(NO_3)_{2\,(aq)} \rightarrow Ba_3(PO_4)_{2\,(s)} + 3NaNO_{3\,(aq)}$$

Barium is next and out of balance. We try tripling the barium nitrate; that would balance with the three bariums in barium phosphate.

$$2Na_3PO_{4 \, (aq)} + 3Ba(NO_3)_{2 \, (aq)} \rightarrow Ba_3(PO_4)_{2 \, (s)} + 3NaNO_{3 \, (aq)}$$

This leaves us with six nitrates on the left and only three on the right. So we double the three sodium nitrates on the right.

$$2Na_3PO_{4 \, (aq)} + 3Ba(NO_3)_{2 \, (aq)} \rightarrow Ba_3(PO_4)_{2 \, (s)} + 6NaNO_{3 \, (aq)}$$

The reaction appears to be balanced. A double check confirms that it is balanced.

The trick with balancing equations is to use a pattern and stick to it. The system presented here is called **TRIAL-AND-ERROR** balancing of equations. This name does tell the story—a trial at balancing is made and is corrected when errors show up. This trial-and-error approach is repeated for each of the components of the chemical reaction until everything is in balance.

NOTE: If you write the reactants and products in the chemical reaction types covered in this chapter and cannot get the equation to balance, it may be that one of the compounds is incorrectly written. Another possibility is that the reaction is an oxidation reduction reaction (redox). Redox reactions do not usually balance using trial and error.

OXIDATION REDUCTION (REDOX)

If a particular atom loses an electron, it becomes a **CATION** (a positively charged particle). That electron has to go somewhere. Suppose there is an atom that gains an electron and becomes an **ANION** (a negatively charge particle). Now we know where the electron went.

The loss of an electron (or electrons) is **OXIDATION**. **REDUCTION** is the gain of an electron (or electrons). Oxidation and reduction reaction sets are often called **REDOX** reactions for convenience. These reactions are paired because of the law of conservation of matter (matter can neither be created nor destroyed). In other words, we have an accounting job to do to make certain that we neither gain matter from anywhere nor lose matter into the great void. We did that when we balanced chemical equations and for the same reason.

Electricity is the flow of electrons. We can use the flow of electrons that occurs during redox reactions to power devices. One way in which we do just that is the use of a battery to operate a flashlight, a radio, or a calculator. Regardless of the work that is being done, we have chemical reactions to balance. The balancing of redox reactions is somewhat different from any other type of reaction because the trial-and-error method we used before will not always work with redox reactions.

The easiest technique of balancing redox reactions for the purposes of this chapter is the half-reaction technique. We identify the oxidation and reduction half reactions that will add together to provide us with the complete redox reaction. Before we do any balancing, let's pin down the terminology that applies to redox reactions using the production of iron II sulfide from the elements.

$$Fe + S \rightarrow FeS$$

The iron atom on the left of the equation has an oxidation number (charge or valence) of zero, as do all atoms. The iron on the right is participating in a chemical compound; its oxidation number is +2. It has lost two e^-; it has been oxidized.

The sulfur on the left of the equation is also an atom; the charge is zero for the same reason as the iron. The sulfur on the right side of the equation is in a chemical compound; it has an oxidation number of –2. The sulfur has gained two electrons and has been reduced.

Furthermore, the electron loss of the iron can be thought of as being caused by the fact that sulfur will gain two electrons. Since the iron is being oxidized, the sulfur is the **OXIDIZING AGENT**. We can say the same thing about the sulfur. The electron gain of the sulfur occurs because the iron is willing to give up two electrons. Since the sulfur is being reduced, the iron is the **REDUCING AGENT** (Table 5.1).

Table 5.1
Redox
terminology.

Fe	+	S	→	FeS
Loses 2e⁻ &		Gains 2e⁻ &		
is oxidized.		is reduced.		
Is reducing agent.		Is oxidizing agent.		

BALANCING REDOX REACTIONS

The half-reaction technique of balancing redox reactions accounts for all of the matter and the overall charge of each side of the equation. The first step is to determine the half reactions involved in the net equation; they describe what happens to anything that changes in oxidation number. The **NET REACTION** is the final equation that is produced when we add the half reactions together.

There are other techniques of balancing redox equations. The half-reaction method is the one that best lends itself to the study of electrochemistry because we are required to recognize oxidation and reduction half reactions later in this chapter.

Let's start with the production of iron II sulfide from the elements and balance by the half-reaction technique, as shown in Example 5.11.

Example 5.11: Using the half-reaction technique, balance the production of iron II sulfide from the elements.

$$Fe + S \rightarrow FeS$$

Solution: The oxidation numbers are determined for each of the participants in the equation. Free atoms (not combined chemically) in the equation have an oxidation number of zero. We notice that the sulfur on the right is in a compound and written in the position that would indicate a negative oxidation number. The only negative oxidation number for sulfur is –2; therefore, the iron in this compound must be a +2.

$$Fe^0 + S^0 \rightarrow Fe^{+2}S^{-2}$$

We have identified the iron and the sulfur as the substances that change in charge. We write the half reactions for each substance that changes in charge and indicate in the reaction the movement of electrons that explains that change.

$$\text{Oxidation: } Fe \rightarrow Fe^{+2} + 2e^-$$
$$\text{Reduction: } 2e^- + S \rightarrow S^{-2}$$

The two half reactions are balanced from the standpoint of the atoms. Notice that the net charge on the left and right of each of the half reactions is the same. The iron half reaction has a net charge of zero on both sides; the sulfur half reaction has a net charge of –2 on both sides. Half reactions must balance not only the atoms but the net charges.

We also notice that there are two electrons on the right of the iron half reaction and two electrons on the left of the sulfur half reaction. We can add half reactions. During the addition of the half reactions, the electrons will cancel; we have the net reaction and, as with all net chemical reactions, there are no electrons indicated. In other words, *the electrons must cancel when we add half reactions to produce the net reaction.*

$$\text{Oxidation:} \qquad Fe \rightarrow Fe^{+2} + \cancel{2e^-}$$
$$\text{Reduction:} \qquad \cancel{2e^-} + S \rightarrow S^{-2}$$
$$\text{Net Reaction:} \quad \overline{Fe + S \rightarrow FeS}$$

The reaction in Example 5.11 can be balanced by the trial-and-error method we used earlier in this manual. The reason for Example 5.11 was to investigate the use of the half-reaction technique so we could balance reactions that were not quite so friendly.

Let's take a look at a more complex–looking reaction in Example 5.12.

Example 5.12: Using the half-reaction technique, balance the reaction of metallic aluminum and the aqueous tin cation indicated.

$$Al_{(s)} + Sn^{+4}_{(aq)} \rightarrow Al^{+3}_{(aq)} + Sn^{+2}_{(aq)}$$

Solution: This reaction does not include any of the **SPECTATOR IONS**; they are the ions that do not participate in the chemical reaction. The spectator ion in this reaction is the anion of the tin compound. It is probably nitrate or some other ion that produces a soluble compound of the tin cation. An equation that does not display the spectator ions is a **NET EQUATION**.

The aluminum changes in oxidation number from zero to +3; this is the oxidation half reaction.

$$Al \rightarrow Al^{+3} + 3e^-$$

The tin cation changes from +4 to +2; this is the reduction half reaction.

$$2e^- + Sn^{+4} \rightarrow Sn^{+2}$$

Notice that the electrons are added to the side of the half reaction that applies. We do not "subtract" matter to balance an equation.

When we add the two half reactions to obtain the net reaction, the electrons must cancel completely. We can cancel electrons only if we have the same number on the two sides of the half reactions. We can multiply an equation through completely without changing the ratio of the participants in the equation. The best procedure is to multiply to obtain the smallest number that contains the common factors 2 and 3 in the case of this reaction. The smallest number is six (2×3 and 3×2).

$$2(Al \rightarrow Al^{+3} + 3e^-) \qquad = \quad 2Al \rightarrow 2Al^{+3} + 6e^-$$
$$3(2e^- + Sn^{+4} \rightarrow Sn^{+2}) \qquad = \quad 6e^- + 3Sn^{+4} \rightarrow 3Sn^{+2}$$

Notice that the electrons do cancel; there are six electrons on the right and the left of the half reactions. We add the two half reactions and have a balanced net reaction.

$$2Al \qquad \rightarrow 2Al^{+3} + \cancel{6e^-}$$
$$\underline{\cancel{6e^-} + 3Sn^{+4} \quad \rightarrow 3Sn^{+2}}$$
$$2Al + 3Sn^{+4} \quad \rightarrow 2Al^{+3} + 3Sn^{+2}$$

Since it is always a good idea to check your answer in any problem, we look to the numbers of each substance to provide a partial check. There are equal numbers of aluminum on the right and left, as there are equal numbers of tin on the right and the left of the net equation. We also look at the overall charge on the left and right of the net equation; it must be the same charge. The charge on the left is +12 from the three Sn^{+4} cations. The charge on the right is also +12 because there is +6 from the aluminum and +6 from the tin cations. This net equation is balanced.

Redox reactions often occur in either acid or base solutions. The acid environment provides hydrogen ions and water with which we can balance equations. The basic solutions allow the use of hydroxide ions and water for balancing. As with the preceding examples, the net equations are used to determine the half reactions. The difference is that there may be elements missing from one side of the half reaction that are on the other side. Of course, we still cannot create matter, nor may we destroy it when balancing equations. Since the missing elements are hydrogen and oxygen, we can use them from the environment (an acid or a base or water).

Example 5.13: Balance the reaction of the Fe^{+2} cation with the permanganate anion in an acid solution.

$$Fe^{+2} + MnO_4^{-1} \rightarrow Fe^{+3} + Mn^{+2}$$

Solution: The first step in the balancing of this reaction is the labeling of each of the participants with its oxidation number. A helpful fact is that oxygen is a –2 under nearly all circumstances in a chemical compound. This helps with determining manganese's oxidation number in the anion.

$$Fe^{+2} + Mn^{+7}O_4^{-8=-1} \rightarrow Fe^{+3} + Mn^{+2}$$

We separate the reaction into the half reactions on the basis of the changes in oxidation number. We place the number of electrons in each half reaction that explains the change in oxidation number.

$$Fe^{+2} \rightarrow Fe^{+3} + 1e^-$$
$$5e^- + MnO_4^{-1} \rightarrow Mn^{+2}$$

The half reaction for the oxidation of iron is balanced.

The half reaction for manganese is not balanced because there is oxygen on the left and not on the right. Additionally, the overall charge on the left is different from on the right (–6 and +2, respectively). The notation that we are *in an acid solution* tells us that we may add H^{+1} ions and H_2O as needed—*we need them*. Let's balance the oxygen by adding four water molecules to the right of the half reaction.

$$5e^- + MnO_4^{-1} \rightarrow Mn^{+2} + 4H_2O$$

The addition of the water is to balance the oxygens, but we have hydrogen on the right and none on the left. Since we are performing this reaction in an acid environment, we can add hydrogen ions wherever they are needed. We need hydrogen on the left. Eight hydrogen ions added to the left will balance the hydrogen.

$$8H^{+1} + 5e^- + MnO_4^{-1} \rightarrow Mn^{+2} + 4H_2O$$

This half appears to be balanced. We can check the balancing by looking at the numbers of atoms on the left and the right. There are eight hydrogens on both sides; there is one manganese on both sides; and there are four oxygens on both sides.

We also must look at the net (overall) charge on both sides of the half reaction. The net charge on the left is +2; this is the same as the net charge on the right of the half reaction. This half reaction for the reduction of manganese is balanced.

We can add the two half reactions to produce the oxidation–reduction reaction.

$$Fe^{+2} \rightarrow Fe^{+3} + 1e^-$$
$$8H^{+1} + 5e^- + MnO_4^{-1} \rightarrow Mn^{+2} + 4H_2O$$

Before we add the half reactions, we must eliminate the electrons from the half reactions since the redox reaction is not written with electrons. We multiply the two equations with whole numbers that result in equal numbers of electrons. The electrons are located on opposite sides of the half reactions and will cancel. Let's multiply the iron half reaction by 5 and leave the manganese reaction as is (essentially, multiply by 1).

$$5[Fe^{+2} \rightarrow Fe^{+3} + 1e^-]$$
$$8H^{+1} + 5e^- + MnO_4^{-1} \rightarrow Mn^{+2} + 4H_2O$$

We cancel the electrons and add the half reactions.

$$5Fe^{+2} \rightarrow 5Fe^{+3} + \cancel{5e^-}$$
$$\underline{8H^{+1} + \cancel{5e^-} + MnO_4^{-1} \rightarrow Mn^{+2} + 4H_2O}$$
$$8H^{+1} + MnO_4^{-1} + 5Fe^{+2} \rightarrow 5Fe^{+3} + Mn^{+2} + 4H_2O$$

This redox reaction appears to be balanced. We already know that the numbers of particles are balanced since the half reactions are balanced. We check to see if the equation is balanced with reference to the total charge on the left (+17) and the total charge on the right (+17). The charges are the same.

The redox reaction is balanced.

Example 5.14: Balance the reduction of chromium from the dichromate ion by chloride ions.

$$Cr_2O_7^{-2} + Cl^{-1} \rightarrow Cr^{+3} + Cl_2$$

Solution: This example tells us about the environment of the system in the equation. The redox reaction proceeds in an acid solution.

Let's separate the redox reaction into its half reactions by first labeling each individual with its oxidation number.

$$Cr_2^{+12}O_7^{-14\,=\,-2} + Cl^{-1} \rightarrow Cr^{+3} + Cl_2$$

We separate the reaction into the half reactions on the basis of the changes in oxidation number.

$$Cr_2^{+12}O_7^{-14\,=\,-2} \rightarrow Cr^{+3}$$
$$Cl^{-1} \rightarrow Cl_2$$

The chromium reduction half reaction is not balanced. Notice that the oxidation number associated with chromium in the dichromate ion is for *two* chromiums. The first step is to balance the participants that change oxidation number. We did not have to consider this step in Example 5.3 because we started with one iron and one manganese on both sides. Let's balance the chromium and place the appropriate oxidation number for each chromium.

$$Cr_2^{+6\,each}O_7^{-2} \rightarrow 2Cr^{+3\,each}$$

The number of electrons that move is due to the *total* oxidation number change. The total oxidation number of the chromium on the left is +12 ($2 \times +6$); the total oxidation number of the chromium on the right is +6 ($2 \times +3$). Six electrons are involved in the reduction of the chromium.

$$6e^- + Cr_2O_7^{-2} \rightarrow 2Cr^{+3}$$

There is oxygen on the left side of the half reaction, but there is no oxygen on the right side. We place water on the right to account for the oxygen on the left. Seven H_2O's are placed on the right to balance the seven oxygens on the left.

$$6e^- + Cr_2O_7^{-2} \rightarrow 2Cr^{+3} + 7H_2O$$

We place hydrogen ions on the left to balance the 14 H's on the right.

$$14H^{+1} + 6e^- + Cr_2O_7^{-2} \rightarrow 2Cr^{+3} + 7H_2O$$

The net charge balances. We now balance the chlorine oxidation half reaction.

$$Cl^{-1} \rightarrow Cl_2$$

We notice that there is one chlorine on the left, but there are two chlorines on the right. We double the chlorine on the left.

$$2Cl^{-1} \rightarrow Cl_2$$

The chlorine on the left is oxidized and has lost a total of two electrons. We account for this loss by placing the electrons into the half reaction. Since we cannot subtract matter, we place the two electrons on the right of the half reaction.

$$2Cl^{-1} \rightarrow Cl_2 + 2e^-$$

We can add the two half reactions after we multiply by factors that will cause the electrons to cancel. In this pair of reactions, we have six electrons from the chromium half reaction and two electrons from the chlorine half reaction. The smallest number containing 6 and 2 as common factors is 6. Therefore, we multiply the chromium reaction by 1 (which is the same as doing nothing) and multiply the chlorine half reaction by 3. Since the electrons are on opposite sides of the reaction, they cancel.

$$14H^{+1} + 6e^- + Cr_2O_7^{-2} \rightarrow 2Cr^{+3} + 7H_2O$$
$$6Cl^{-1} \rightarrow 3Cl_2 + 6e^-$$
$$\overline{14H^{+1} + Cr_2O_7^{-2} + 6Cl^{-1} \rightarrow 2Cr^{+3} + 7H_2O + 3Cl_2}$$

A check of this redox reaction tells us that it is balanced. The particles balance and so do the net charges (+6) from the two sides of the equation.

The same techniques as shown in Examples 5.13 and 5.14 are used to balance redox reactions in a basic solution. The only difference is that we can place hydroxide ions and/or water wherever needed; they both contain oxygen. Let's look at Example 5.15 for the trick used to work with added oxygen from two sources (OH^{-1} and H_2O).

Example 5–15: Balance the following oxidation–reduction reaction. Indicate which element is oxidized and which element is reduced.

$$S_2O_3^{-2} + I_2 \rightarrow SO_4^{-2} + I^{-1}$$

Solution: Since oxygen rarely changes oxidation number, the half reactions are the sulfur reaction and the iodine reaction. We write the two half reactions and balance the particles that we can (the sulfur and the iodine).

$$S_2O_3^{-2} \rightarrow 2SO_4^{-2}$$
$$I_2 \rightarrow 2I^{-1}$$

Let's balance the sulfur half reaction first. The oxidation numbers for the sulfur on both sides are written in so we can determine the number of electrons that move to or from the sulfur on the left.

$$S_2^{+4}O_3^{-6\,=\,-2} \rightarrow 2S^{+6}O_4^{-8\,=\,-2}$$

The total charge of the sulfur on the left of the half reaction is +4; the total charge of the sulfur on the right of the half reaction is +12 (two sulfurs). We account for eight electrons because of the total change in charge. This is an oxidation half reaction because sulfur is losing electrons (gaining in charge).

$$S_2O_3^{-2} \rightarrow 2SO_4^{-2} + 8e^-$$

Our next consideration is to balance the oxygens. We can add either water or hydroxide ions, but which one and where? If we place the water, as we did in the previous examples, are we certain it is on the correct side? Since the equation has to balance chargewise, let's place the charged choice, OH^{-1}, on the left. We use enough hydroxide to balance the charges.

$$10OH^{-1} + S_2O_3^{-2} \rightarrow 2SO_4^{-2} + 8e^-$$

The net charges are the same on both sides, -12. Now we have hydrogen on the left, and there is none on the right. We need 10 hydrogens, so let's use five waters.

$$10OH^{-1} + S_2O_3^{-2} \rightarrow 2SO_4^{-2} + 8e^- + 5H_2O$$

We balance the number of iodines and place the electrons to balance charges.

$$2e^{-1} + I_2 \rightarrow 2I^{-1}$$

Notice that this is a reduction half reaction because the iodine is gaining electrons (or is reducing in charge).

We can add the balanced half reactions after we multiply through each half reaction to cancel the electrons. The lowest number with the common factors of 8 and 2 is 8. Let's multiply the sulfur reaction by 1 and the iodine reaction by 4, and then add the half reactions.

$$10OH^{-1} + S_2O_3^{-2} \rightarrow 2SO_4^{-2} + \cancel{8e^-} + 5H_2O$$
$$\underline{\cancel{8e^{-1}} + 4I_2 \rightarrow 8I^{-1}}$$
$$10OH^{-1} + S_2O_3^{-2} + 4I_2 \rightarrow 2SO_4^{-2} + 5H_2O + 8I^{-1}$$

A check of the balancing of the equation tells us that the particles are in balance. The charge on the left is the same as the net charge on the right. This oxidation–reduction equation is balanced.

QUESTIONS AND PROBLEMS

Balance the following equations:

1. ___ Li + ___ S → ___ ___

2. ___ Mg + ___ I$_2$ → ___ ___

3. ___ B + ___ Cl$_2$ → ___ ___

4. ___ Zn + P → ___ ___

5. ___ Ba + C → ___ ___

6. ___ N$_2$ + ___ O$_2$ → ___ N$_2$O$_3$

7. ___ Cu + Br$_2$ → ___ CuBr

8. ___ Fe + ___ S → ___ Fe$_2$S$_3$

9. ___ S + O$_2$ → ___ SO$_3$

10. ___ Hg + ___ Br$_2$ → ___ HgBr

11. Aluminum reacts with oxygen to yield aluminum oxide.

12. Beryllium reacts with fluorine to produce beryllium fluoride.

13. The reaction of silicon and oxygen produces silicon dioxide (sand).

14. Nickel reacts with sulfur to yield nickel II sulfide.

15. Calcium reacts with carbon to produce calcium carbide.

16. Write the reaction for the production of calcium carbonate from the elements.

17. What is the balanced reaction for the production of sulfuric acid from the elements?

18. What is the reaction when oxygen and sulfur react?

19. Write the balanced chemical reaction(s) for the reaction of nitrogen and oxygen.

20. ___ Na + ___ HCl → ___ NaCl + ___ H$_2$

21. ___ Mg + ___ H$_2$SO$_4$ → ___ MgSO$_4$ + ___ H$_2$

22. ___ Al + ___ H$_2$SO$_4$ → ___ Al$_2$(SO$_4$)$_3$ + ___ H$_2$

23. ___ Ra + ___ H$_3$PO$_4$ → ___ Ra$_3$(PO$_4$)$_2$ + ___ H$_2$

24. ___ Na + ___ CCl$_4$ → ___ NaCl + ___ C

25. ___ Ca + ___ AlBr$_3$ → ___ CaBr$_2$ + ___ Al

26. ___ K + ___ SO$_3$ → ___ K$_2$O + ___ S

27. ___ Fe$_2$S$_3$ + ___ O$_2$ → ___ Fe$_2$O$_3$ + ___ S

28. ___ Na + ___ BPO$_4$ → ___ ___ + ___ ___

29. ___ Fr + ___ CuCO$_3$ → ___ ___ + ___ ___

30. ___ Ag$_2$S + ___ F$_2$ → ___ ___ + ___ ___

31. ___ AgF + ___ O$_2$ → ___ ___ + ___ ___

32. ___ Sr + ___ CaCO$_3$ → ___ ___ + ___ ___

33. What is the balanced equation that describes the reaction of iron with sulfuric acid to produce iron III sulfate?

34. Write the balanced reaction between copper and acetic acid; copper II acetate is produced.

35. Will radium oxide and fluorine react? If so, write the equation.

36. Write the reaction between sodium chloride and neon.

37. ___ H_2SO_4 + ___ $Be(OH)_2$ → ___ ___ + ___ ___

38. ___ HCl + ___ $Al(OH)_3$ → ___ ___ + ___ ___

39. ___ H_3PO_4 + ___ $Zn(OH)_2$ → ___ ___ + ___ ___

40. ___ $HC_2H_3O_2$ + ___ $Cu(OH)_2$ → ___ ___ + ___ ___

41. ___ H_2CO_3 + ___ $U(OH)_5$ → ___ ___ + ___ ___

42. ___ $AgNO_{3\,(aq)}$ + ___ $MgCl_{2\,(aq)}$ → ___ $AgCl_{(s)}$ + ___ $Mg(NO_3)_{2\,(aq)}$

43. ___ $HCl_{(aq)}$ + ___ $CaCO_{3\,(s)}$ → ___ $CaCl_{2\,(aq)}$ + ___ $CO_{2\,(g)}$ + ___ $H_2O_{(l)}$

44. ___ $Pb(NO_3)_{2\,(aq)}$ + ___ $Al_2(SO_4)_{3\,(aq)}$ → ___ $PbSO_{4\,(s)}$ + ___ $Al(NO_3)_{3\,(aq)}$

45. ___ $MgCl_{2\,(aq)}$ + ___ $NaOH_{(aq)}$ → ___ $Mg(OH)_{2\,(s)}$ + ___ $NaCl_{(aq)}$

Balance the following redox reactions:

46. $Zn_{(s)} + Cu^{+2}_{(aq)} \rightarrow Zn^{+2}_{(aq)} + Cu_{(s)}$

47. $Ni^{+2}_{(aq)} + Cd_{(s)} \rightarrow Ni_{(s)} + Cd^{+2}_{(aq)}$

48. $Fe_{(s)} + Ag^{+1}_{(aq)} \rightarrow Fe^{+3}_{(aq)} + Ag_{(s)}$

49. $Al_{(s)} + Mg^{+2}_{(aq)} \rightarrow Al^{+3}_{(aq)} + Mg_{(s)}$

50. $Cu_{(s)} + Ca^{+2}_{(aq)} \rightarrow Cu^{+1}_{(aq)} + Ca_{(s)}$

51. $Ba_{(s)} + As^{+3}_{(aq)} \rightarrow Ba^{+2}_{(aq)} + As_{(s)}$

52. $Sr_{(s)} + Al^{+3}_{(aq)} \rightarrow Sr^{+2}_{(aq)} + Al_{(s)}$

53. $NO_3^{-1} + S^{-2} \xrightarrow{acid} NO + S$

54. $MnO_4^{-1} + Cl^{-1} \xrightarrow{acid} Mn^{+2} + Cl_2$

55. $Hg_2S + I_2 \xrightarrow{acid} HgI + S$

56. $Pb + NO_3^{-1} \xrightarrow{acid} PbO_2 + NO_2$

57. $Se_{(s)} + SnO_2^{-2}{}_{(aq)} \xrightarrow{base} SeO_3^{-2}{}_{(aq)} + Sn_{(s)}$

58. $Ba^{+2} + N^{-2} \xrightarrow{base} Ba + NO_2^{-1}$

59. $Se + ZnO_2^{-2} \xrightarrow{base} SeO_3^{-2} + Zn$

60. $S + Re \xrightarrow{base} HS^{-1} + ReO_4^{-1}$

61. Given the reaction $Zn_{(s)} + Cu^{+2}{}_{(aq)} \rightarrow Zn^{+2}{}_{(aq)} + Cu_{(s)}$
    Identify the element that is oxidized.

62. Given the reaction $Sr_{(s)} + Al^{+3}{}_{(aq)} \rightarrow Sr^{+2}{}_{(aq)} + Al_{(s)}$
    Identify the element that is oxidized.

63. Given the reaction $NO_3^{-1} + S^{-2} \rightarrow NO + S$
    Identify the element that is oxidized.

64. Given the reaction (basic) $Fe(OH)_2 + MnO_4^{-1} \rightarrow Fe(OH)_3 + MnO_2$
    Identify the element that is oxidized.

65. Given the reaction $Zn_{(s)} + Cu^{+2}{}_{(aq)} \rightarrow Zn^{+2}{}_{(aq)} + Cu_{(s)}$
    Identify the element that is reduced.

66. Given the reaction $Sr_{(s)} + Al^{+3}{}_{(aq)} \rightarrow Sr^{+2}{}_{(aq)} + Al_{(s)}$
    Identify the element that is reduced.

67. Given the reaction $NO_3^{-1} + S^{-2} \rightarrow NO + S$
    Identify the element that is reduced.

68. Given the reaction (basic) $Fe(OH)_2 + MnO_4^{-1} \rightarrow Fe(OH)_3 + MnO_2$
    Identify the element that is reduced.

69. Given the reaction $Zn_{(s)} + Cu^{+2}{}_{(aq)} \rightarrow Zn^{+2}{}_{(aq)} + Cu_{(s)}$
    Identify the element that is the oxidizing agent.

70. Given the reaction $Sr_{(s)} + Al^{+3}{}_{(aq)} \rightarrow Sr^{+2}{}_{(aq)} + Al_{(s)}$
    Identify the element that is the oxidizing agent.

71. Given the reaction $NO_3^{-1} + S^{-2} \rightarrow NO + S$
    Identify the element that is the oxidizing agent.

72. Given the reaction (basic) $Fe(OH)_2 + MnO_4^{-1} \rightarrow Fe(OH)_3 + MnO_2$
    Identify the element that is the oxidizing agent.

73. Given the reaction $Zn_{(s)} + Cu^{+2}_{(aq)} \rightarrow Zn^{+2}_{(aq)} + Cu_{(s)}$
    Identify the element that is the reducing agent.

74. Given the reaction $Sr_{(s)} + Al^{+3}_{(aq)} \rightarrow Sr^{+2}_{(aq)} + Al_{(s)}$
    Identify the element that is the reducing agent.

75. Given the reaction $NO_3^{-1} + S^{-2} \rightarrow NO + S$
    Identify the element that is the reducing agent.

76. Given the reaction (basic) $Fe(OH)_2 + MnO_4^{-1} \rightarrow Fe(OH)_3 + MnO_2$
    Identify the element that is the reducing agent.

CALCULATIONS AND CHEMICAL EQUATIONS

• Mass Calculations • Limiting Reactant •
• Questions and Problems •

MASS CALCULATIONS

We know that chemicals react in specific ways requiring definite amounts to satisfy the law of conservation of matter. We express mass conservation in the form of a balanced chemical equation.

If we know how much of one of the participants in a chemical reaction is present, we can calculate the amount of the others. The calculation may be performed in moles or in masses. This concept is important to industry, chemical manufacturing, hazardous materials clean-up teams, and dosage calculations in health care.

Gaseous hydrogen chloride and solid calcium hydroxide react to produce harmless calcium chloride and water. Hydrogen chloride is a very dangerous gas to inhale; it is extremely corrosive; severe damage to the lungs will occur. How would we go about determining how much calcium hydroxide is needed to neutralize a specific amount of hydrogen chloride? Take a look at Example 6.1.

Example 6.1: What is the number of moles of solid calcium hydroxide that would just react with 6 moles of hydrogen chloride gas?

Solution: Since *react* is used in this problem, we write a balanced chemical reaction.

$$Ca(OH)_2 + 2HCl \rightarrow CaCl_2 + 2HOH$$

The balanced equation is used to present both the equation and given information back to us so that we may calculate the answer.

Equation: **1 mol**   **2 mol**
$$Ca(OH)_2 + 2HCl \rightarrow CaCl_2 + 2HOH$$
Given: y **6 mol**

We set up the ratio and proportion from the given information and the equation information.

$$\frac{1 \text{ mol Ca(OH)}_2}{y} = \frac{2 \text{ mol HCl}}{6 \text{ mol HCl}}$$

Cross multiplication is the quickest way to resolve the equation to a linear arrangement.

$$\frac{1 \text{ mol Ca(OH)}_2}{y} \diagdown\!\!\!\!\times \frac{2 \text{ mol HCl}}{6 \text{ mol HCl}}$$

The equation becomes

$$2 \text{ mol HCl} \times y = 1 \text{ mol Ca(OH)}_2 \times 6 \text{ mol HCl}$$

We divide both sides by 2 mol HCl in order to isolate the y.

$$y = \frac{1 \text{ mol Ca(OH)}_2 \times 6 \cancel{\text{ mol HCl}}}{2 \cancel{\text{ mol HCl}}}$$

Notice that the mol HCl cancel out, leaving us with $y = $ mol Ca(OH)$_2$, which is the answer in units we need. We divide and arrive at the final answer.

$$y = 3 \text{ mol Ca(OH)}_2$$

*NOTE: A y is used as the symbol for the unknown quantity rather than
the traditional x. The × is used to indicate multiplication; y is used for
the unknown symbol to avoid confusion.*

Calculations performed in moles are not the complete answer when dealing
with chemical reactions. The problem is that balances weigh masses, not moles.
Solving problems in mass units involves a conversion from the moles to the desired
units of mass (usually grams).

Example 6.2: What weight of calcium hydroxide is necessary to neutralize 45.0
grams of hydrogen chloride?

Solution: Notice that this problem is very similar to Example 6.1 except that we are
dealing with grams instead of moles. Also, *neutralize* indicates that there is a
chemical reaction; this equation is a metathesis (double replacement) reaction.

$$Ca(OH)_2 + 2HCl \rightarrow CaCl_2 + 2HOH$$

We write in the equation information and given information.

Equation: **1 mol**   **2 mol**
$$Ca(OH)_2 + 2HCl \rightarrow CaCl_2 + 2HOH$$
Given: y **45.0 g**

We cannot set up a ratio and proportion because we have moles and grams; we
need to be in the same units. Convert moles to grams using the gram molecular weight
and a little dimensional analysis. Continue to solve for the answer.

$$\text{For } Ca(OH)_2: \ 1 \text{ mol } Ca(OH)_2 \times \frac{74.0926 \text{ g } Ca(OH)_2}{1 \text{ mol } Ca(OH)_2} = 74.0926 \text{ g } Ca(OH)_2$$

$$\text{For HCl:} \quad 2 \text{ mol HCl} \times \frac{36.4606 \text{ g HCl}}{1 \text{ mol HCl}} = 77.9212 \text{ g HCl}$$

We replace the mole information with gram information.

Equation: **74.0926 g   72.9212 g**
$$Ca(OH)_2 + 2HCl \rightarrow CaCl_2 + 2HOH$$
Given: y **45.0 g**

The ratio and proportion is written; cross multiplication is indicated.

$$\frac{74.0926 \text{ g } Ca(OH)_2}{y} \diagtimes \frac{72.9212 \text{ g HCl}}{45.0 \text{ g HCl}}$$

The cross multiplication gives us a linear equation.

$$72.9212 \text{ g HCl} \times y = 74.0926 \text{ g } Ca(OH)_2 \times 45.0 \text{ g HCl}$$

We divide to isolate the unknown and look to see that the appropriate answer
is given in units. This solution in units tells us that we have set up and solved correctly
to this point.

$$y = 74.0926 \text{ g } Ca(OH)_2 \times \frac{45.0 \text{ g HCl}}{72.9212 \text{ g HCl}} = 45.7 \text{ g } Ca(OH)_2$$

---

*NOTE: Do not round off until the last step of the problem. If you do
round off earlier, you introduce error that compounds (is multiplied)
throughout the problem. The complete weights are used from the
Periodic Table; rounding to the appropriate number of digits is the
last step.*

Example 6.3:   0.752 grams of iron react with an excess of sulfur producing iron III sulfide. How much iron III sulfide is produced by the reaction?

Solution: A balanced chemical reaction is called for again.
$$2Fe + 3S \rightarrow Fe_2S_3$$
We record the problem information into the equation. We also note the information that comes directly from the equation data.

Equation: 2 mol   1 mol
$$2Fe + 3S \rightarrow Fe_2S_3$$
Given: 0.752 g y

In Example 6.2, we calculated a separate conversion from moles to grams by means of gram molecular weight. Let's set up this problem as a single calculation.

Equation: 2 ~~mol~~ × 55.874 $^g/_{mol}$ 1 ~~mol~~ × 207.946 $^g/_{mol}$
$$2Fe + 3S \rightarrow Fe_2S_3$$
Given: 0.752 g y

Notice that the mol associated with the iron cancels; also, the mol from the Fe_2S_3 cancels. In essence, this is a shortcut technique.

We set up the ratio and proportion.

$$\frac{2 \times 55.874 \text{ g Fe}}{0.752 \text{ g Fe}} \bowtie \frac{1 \times 207.946 \text{ g Fe}_2S_3}{y}$$

The next step is cross multiplication.
$$2 \times 55.874 \text{ g Fe} \times y = 1 \times 207.946 \text{ g Fe}_2S_3 \times 0.752 \text{ g Fe}$$
We divide both sides by 2 × 55.874 g Fe in order to isolate y.

$$y = \frac{1 \times 207.946 \text{ g Fe}_2S_3 \times 0.752 \text{ g Fe}}{2 \times 55.874 \text{ g Fe}}$$

The g Fe cancel out. We are left with $y = $ g Fe_2S_3, the answer (units) we were asked to calculate.

$$y = 1.39936 \text{ g Fe}_2S_3 = 1.40 \text{ g Fe}_2S_3 \text{ produced}$$

NOTE: The setup technique used in Example 6.3 is to our advantage for two reasons. The first is that we may cancel at the original setup and do not have to carry more complex units than are necessary. Second, all our calculations are performed after the algebraic manipulations; there is less chance for calculation error.

LIMITING REACTANT

Some problems contain information for all of the reactants. The masses often are not the appropriate amounts to react completely. If there is a reactant that is in short supply, the reaction stops when that reactant is completely used. The reactant that stops the reaction (is used up first) is called the **LIMITING REACTANT** because it limits how far the reaction will proceed.

HINT: If a problem gives the information in either moles or mass for all of the reactants, more than likely you will be dealing with a limiting reactant problem. It pays to take the time and test for a limiting reactant.

Example 6.4: 30.0 grams methane react with 20.0 grams oxygen. Calculate the mass of carbon dioxide produced by the reaction

$$CH_4 + 2O_2 \rightarrow CO_2 + 2H_2O$$

Solution: We write the equation and given information into the equation. However, first we have to convert the information so it is all in the same units. Grams are the unit of choice; we usually weigh on lab balances that express weights in grams.

For the methane: $\quad 1 \text{ mol CH}_4 \times \dfrac{16.0426 \text{ g CH}_4}{1 \text{ mol CH}_4} = 16.0426 \text{ g CH}_4$

For the oxygen: $\quad 2 \text{ mol O}_2 \times \dfrac{31.998 \text{ g O}_2}{1 \text{ mol O}_2} = 63.9976 \text{ g O}_2$

For the carbon dioxide: $\quad 1 \text{ mol CO}_2 \times \dfrac{44.0098 \text{ g CO}_2}{1 \text{ mol CO}_2} = 44.0098 \text{ g CO}_2$

Writing the gram information into the equation gives us

Equation: $\quad$ 16.0426 g $\quad$ 63.9976 g $\quad\quad$ 44.0098 g
$$CH_4 \quad + \quad 2O_2 \quad \rightarrow \quad CO_2 \quad + \quad 2H_2O$$
Given: $\quad\quad$ 30 g $\quad\quad\quad$ 20.0 g $\quad\quad\quad$ y

It is possible that there will be one reactant that will be used up before the other. We need to test the given information to determine *if* one of the reactants is a limiting reactant. Let's use *t* as the symbol used for the gram test of methane used if all of the oxygen is reacted (*we may use any symbol we wish*). Also, we can use whole numbers for the molecular weights since we only need an approximation.

Equation: $\quad\quad$ 16 g $\quad\quad$ 64 g
$$CH_4 + 2O_2 \rightarrow CO_2 + 2H_2O$$
Test: $\quad\quad\quad\quad$ *t* $\quad\quad$ 20 g

The test ratio and proportion is solved to tell us whether or not there is enough methane to perform the chemical reaction. If there is not enough methane, it is the limiting reactant. If we have more methane than we need, the oxygen is the limiting reactant.

$$\frac{16 \text{ g CH}_4}{t} \neq \frac{64 \text{ g O}_2}{20 \text{ g O}_2}$$

The cross multiplication yields

$$64 \text{ g O}_2 \times t = 16 \text{ g CH}_4 \times 20 \text{ g O}_2$$

Dividing both sides by 32 g O_2 and finishing the calculation finishes the test.

$$t = \frac{16 \text{ g CH}_4 \times 20 \text{ g O}_2}{64 \text{ g O}_2} = 5 \text{ g CH}_4$$

This test tells us that only 5 grams of methane will be used in the reaction with 20 grams of oxygen.

Since the reaction stops when all the oxygen is reacted and there is an excess of methane, the limiting reactant is oxygen. The calculations requested by the problem are based on the limiting reactant, oxygen. We do not have to calculate any further on the basis of methane unless we are asked for the amount of the excess.

We are no longer interested in the methane and water. Our calculations now depend on the portion of the equation that applies: $2O_2 \approx CO_2$.

We record the information we have into the balanced equation fragment.

Equation: **63.9976 g**   **44.0098 g**
$$2O_2 \quad \approx \quad CO_2$$
Given: **20.0 g** y

This relationship provides us with the ratio and proportion.

$$\frac{63.9976 \text{ g CO}_2}{20.0 \text{ g O}_2} \diagdown\!\!\!\!\diagup \frac{44.0098 \text{ g CO}_2}{y}$$

The cross multiplication gives us the linear expression of the equation.

$$63.9976 \text{ g O}_2 \times y = 20.0 \text{ g O}_2 \times 44.0098 \text{ g CO}_2$$

We divide both sides by 63.9976 g O_2 and solve for the answer.

$$y = 20.0 \text{ g O}_2 \times \frac{44.0098 \text{ g CO}_2}{63.9976 \text{ g O}_2} = 13.754 \text{ g CO}_2 = 13.8 \text{ g CO}_2$$

Example 6.5: Potassium manganate reacts with hydrochloric acid.
$$3K_2MnO_4 + 4HCl \rightarrow MnO_2 + 2KMnO_4 + 4KCl + 2H_2O$$
How much manganese dioxide is produced by the reaction of 0.25 g potassium manganate with 0.65 g HCl?

Solution: The weights of potassium manganate and hydrogen chloride are given to us. We suspect that there might be a limiting reactant. We test to determine which of the reactants is in short supply. We record data and consider using whole numbers for the molecular weights to run a test using t as the quantity we are testing.

Equation: **3 mol × 197 $^g/_{mol}$**   **4 mol × 36 $^g/_{mole}$**
$$3K_2MnO_4 \quad \approx \quad 4HCl$$
Test: t **0.65 g**

The ratio and proportion we write is

$$\frac{3 \times 197 \text{ g K}_2MnO_4}{t} \diagdown\!\!\!\!\diagup \frac{4 \times 36 \text{ g HCl}}{0.65 \text{ g HCl}}$$

Cross multiplication gives us a linear equation.
$$4 \times 36 \text{ g HCl} \times t = 3 \times 197 \text{ g K}_2MnO_4 \times 0.65 \text{ g HCl}$$
We isolate t by dividing both sides by 4×36 g HCl.

$$t = \frac{3 \times 197 \text{ g K}_2MnO_4 \times 0.65 \text{ g HCl}}{4 \times 36 \text{ g HCl}}$$

Since the units do work out the way we need them, we may calculate the answer.
$$t = 2.67 \text{ g K}_2MnO_4 \text{ required to react with 0.65 g HCl}$$
Our interpretation of these data is that we do not have enough potassium manganate to react with all of the hydrochloric acid. Potassium manganate is a limiting reactant. Our calculations leading to the manganese dioxide produced are based on K_2MnO_4, not HCl. We calculate based on $3K_2MnO_4 \approx MnO_2$.

We write the relationship necessary with the information we have from the problem statement. Complete weights from the Periodic Table are used.

Equation: **3 mol × 197.1322 $^g/_{mol}$**   **1 mol × 86.9368 $^g/_{mol}$**
$$3K_2MnO_4 \quad \approx \quad MnO_2$$
Given: **0.25 g** y

The ratio and proportion is set up.

$$\frac{3 \times 197.1322 \text{ g } K_2MnO_4}{0.25 \text{ g } K_2MnO_4} = \frac{1 \times 86.9368 \text{ } MnO_2}{y}$$

Our cross multiplication gives us

$$3 \times 197.1322 \times y = 0.25 \times 1 \times 86.9368 \text{ g } MnO_2$$

We isolate the y by dividing both sides by 3×197.1322.

$$y = \frac{0.25 \times 1 \times 86.9368 \text{ g } MnO_2}{3 \times 197.1322}$$

We see that the units, g MnO_2, are required by the problem; we may finish the calculation of the answer.

$$y = 0.0368 \text{ g } MnO_2 = 0.04 \text{ g } MnO_2 \text{ produced}$$

NOTE: The calculations of this chapter are often used to predict the yield of a chemical reaction. The mathematical prediction of yield is called the **THEORETICAL YIELD**. *Many times, a chemical reaction will not actually produce the amount of product calculated; the amount that you get from a specific chemical reaction is the* **ACTUAL YIELD**. *These two yields may be compared as the* **PERCENT YIELD** *by a calculation as follows:*

$$Percent \; Yield = \frac{Actual \; Yield}{Theoretical \; Yield} \times 100$$

QUESTIONS AND PROBLEMS

1. Tarnish on silverware can be silver sulfide, Ag_2S. How many moles of sulfur react with 7 moles of silver?

2. One of the components of rust is iron III oxide. What is the number of moles iron that will use up 23 moles oxygen?

3. How many moles of acetic acid are necessary to just react with 4 moles of lithium?

4. Three moles of HCl react with sufficient $Ra(OH)_2$ to use up all the HCl. How much radium hydroxide is required by this reaction?

5. Phosphoric acid reacts with magnesium hydroxide. How many moles of phosphoric acid are necessary to neutralize 13 moles of magnesium hydroxide?

6. The reaction of sodium and sulfuric acid is violent. A great deal of heat is produced and may ignite the hydrogen gas released. How many moles of sulfuric acid are required to produce 5 moles of sodium sulfate?

7. How much sodium is required to produce 50 moles of sodium sulfate from the reaction of sodium and aluminum sulfate?

8. In the upper atmosphere, a reaction producing ozone from oxygen is driven by ultraviolet light, $3O_2 \rightarrow 2O_3$. We can duplicate the conditions of the upper

atmosphere in the lab and start with 0.15 mol O_2. What is our prediction of the amount of O_3 that will be produced?

9. How much copper is required in reaction with 3 g O_2 to produce CuO?

10. What is the weight of zinc necessary to react with 5.00 grams of fluorine?

11. Aluminum and sulfur react. How much sulfur will just react with 22.40 grams of aluminum?

12. Chromium reacts with iodine; one of the possible products is chromium III iodide, CrI_3. Assuming that the production of the compound is from the elements, what is the weight of chromium that reacts with 1.25 grams of iodine?

13. Arsenic V cyanide can be produced by the reaction of 12.0 g As and how much HCN?

14. One of the reactions that occurs in the operation of an automotive battery involves lead reacting with sulfuric acid. How much H_2SO_4 will react completely with 0.25 g Pb? NOTE: The phrase *"will react completely"* tells us that the highest oxidation number is used.

15. How much magnesium is required to react with 0.115 g nitrogen?

16. What is the weight of the copper needed to produce 2.9 g CuS from the reaction of copper and sulfur?

17. How much thorium will react to produce 0.0040 g ThO_2 from the elements?

18. What is the weight of oxygen that will react with elemental phosphorous to produce 0.055 g P_2O_5?

19. An acid/base reaction is used to produce beryllium sulfate. How much acid will react to produce 21 g beryllium sulfate?

20. How much boron acetate is produced by the reaction of 5.50 grams of boron and an excess of acetic acid?

21. What is the amount of PCl_3 that is produced from the reaction of 18.75 grams phosphorous and a sufficient weight of chlorine?

22. What is the theoretical yield of the chemical reaction of 0.155 g Ca and 0.332 g P?

23. Eight grams of sulfur and 7.50 grams of zinc are placed in a reaction vessel. Predict the theoretical yield (the amount produced on reaction).

24. The reaction of 10.00 g Na and 10.00 g S is set up in the lab. What is the

predicted weight of sodium sulfide to be produced?

25. Predict the yield from the reaction of 18.7 g N and 41.5 g oxygen. The product of the reaction is NO_2.

26. What is the weight of ammonium phosphate produced from 0.0255 g ammonium hydroxide and 0.0671 g phosphoric acid?

27. Silver nitrate reacts with sodium chloride to produce the insoluble silver chloride. How much silver chloride is produced when 150 g $AgNO_3$ and 150 g NaCl react?

28. Calculate the weight of iron released when 25.00 g Fe_3O_4 react with 22.25 g CO as follows:

$$Fe_3O_4 + 4CO \rightarrow 3Fe + 4CO_2$$

29. Barium sulfate precipitates when 5.00 g barium nitrate and 5.00 g sodium sulfate react. How much barium sulfate is produced?

30. How many grams of sulfur trioxide are produced when 43.00 g sulfur dioxide react with 37.50 grams oxygen?

GAS LAWS

• Kinetic Molecular Theory • Ideal Gas Law • Combined Gas Law •
• Boyle's Law • Charles' Law • Gay-Lussac's Law •
• Dalton's Law of Partial Pressure • Avogadro's Principle • Graham's Law •
• Questions and Problems •

Gases are interesting—they take the shape of the container, assume the volume of the container, and can be compressed with ease. Additionally, gases are of very low density when compared to solids or liquids. Gases tend to mix well. The atmosphere is a mixture of gases. Dry air is composed primarily of nitrogen and oxygen (a little less than 80% and 20%) with many more gases in low concentration (less than 1%). "Dry air" is mentioned because the water content in the air is variable. The relative humidity portion of a weather report will give you an idea of how much the water content in the air can change. (Relative humidity is the measure of how much water vapor the air *is* holding compared to how much it *could* hold at a specific temperature.)

KINETIC MOLECULAR THEORY

1. Gases are composed of molecules. (Group VIIIA, the inert gases, are considered to be monatomic molecules composed of one atom.)

2. Distances between molecules are very large when compared to the size of the molecules. Because of the large distances, there is very little in the way of attractive forces between molecules.

3. Gas molecules are in constant motion. That motion is in a straight line until collisions with other molecules occur (other gas molecules, the molecules of the container, etc.). No kinetic energy is lost during collisions, and other molecules simply bounce off of each other (perfectly elastic).

4. The speed of the movement of gas molecules is proportional to the temperature in the Kelvin scale (K). The higher the temperature, the faster the movement, and the lower the temperature, the slower the movement.

5. All gases, regardless of composition, have the same average kinetic energy at the same temperature.

IDEAL GAS LAW

The ideal gas law was derived from Boyle's law and Charles' law which will be discussed later. Application of the ideal gas law is to predict the behavior of a gas under a *single set of conditions* (temperature, pressure, and volume).

$$PV = nRT$$

P is the pressure in atmospheres (1 atm = 14.7 lb/in^2 = 760 torr = mm Hg)
V is the volume in liters
n is the number of moles of gas   (moles = grams /molecular weight)
R is the gas constant (0.0821 liter atm/K mole)
T is the temperature in the Kelvin scale   (K = °C + 273)

One application of the ideal gas law predicts the behavior of ideal gases. This means that all gases that are ideal will behave the same way under the same conditions.

Example 7.1: A prediction of the ideal gas law is that the volume of equal numbers of moles of a gases under **STANDARD CONDITIONS** (STP) is the same. *STP is defined as the temperature of 0°C and one atmosphere of pressure.* Assume you have one mole of carbon dioxide, 44 g. What will be the volume when at STP?

Solution: Using the ideal gas law and substituting in the values in the problem statement will give you the answer because all but one of the variables are given in the problem. We must be careful to use the appropriate units. Often, these problems include a trap, such as the temperature's units.

The problî statement tells you that the temperature is °0 C, but the formula calls for temperature to be in Kelvin.

$$K = °C + 273$$
$$K = 0°C + 273$$
$$K = 273$$

We write the formula that applies to the problem.
$$PV = nRT$$

We substitute into the formula and cancel units.

$$1 \text{ atm} \times V = 1 \text{ mol} \times 0.0821 \frac{L \text{ atm}}{\text{mol K}} \times 273K$$

We divide both sides by 1 atm to get the *V* to stand alone and calculate.

$$\frac{1 \text{ atm} \times V}{1 \text{ atm}} = \frac{1 \times 0.0821 \text{ L atm} \times 273}{1 \text{ atm}}$$

$$V = 0.0821 \text{ L} \times 273 = 22.4133 \text{ L} = 22.4 \text{ L}$$

Notice that the units for the unknown, *V*, come out in liters. This is exactly what you want because *V* is volume and must be in liters for the ideal gas law.

Then one mole of any gas conforming to the ideal gas law would occupy 22.4 liters of volume at 0°C (273K) and one atomsphere of pressure by the predictions of the ideal gas law. The words *predict* or *prediction* are associated with many theories and laws, including this one, because science acknowledges that the laws of science do have exceptions. This is why scientists rarely use the words always and never. In the case of the ideal gas law, the calculations of the law vary with the nature of the specific gas molecule and high concentrations. Let's assume for this chapter that the gases involved are ideal gases and, since the concentrations of gases used in most calculations are relatively small, the predictions are pretty good and most gases will conform to the predictions of the ideal gas law.

Example 7.2: How large a container would be required under STP to hold 20.0 grams of oxygen?

Before the calculations of the ideal gas law are applied, we must recall that oxygen is a gas that is composed of diatomic molecules; therefore, the molecular weight of oxygen is two times the weight indicated by the Periodic Table of Elements ($2 \times 15.9994 = 31.9988$ g/mole).

Solution: We write the formula required.

$$PV = nRT$$

We substitute into the formula using V as the symbol for the unknown quantity, volume.

$$1 \text{ atm} \times V = \frac{20 \text{ g}}{31.9988 \text{ g}/\text{mol}} \times 0.0821 \text{ L atm}/\text{mol K} \times 273 \text{ K}$$

We divide out the fraction to give us a linear equation.

$$1 \text{ atm} \times V = 0.625 \times 0.0821 \text{ L atm} \times 273$$

We divide both sides by 1 atm to isolate the V and see V = Liters.

$$\frac{1 \text{ atm} \times V}{1 \text{ atm}} = \frac{0.625 \times 0.0821 \text{ L atm} \times 273}{1 \text{ atm}}$$

We calculate the answer.

$$V = 14.0083 \text{ L} = 14.0 \text{ L}$$

Of course, storing this small amount of oxygen in a 14.0-liter container (approximately $3^1/_2$ gallons) would not be practical. The predictions required to determine a practical-sized container may be performed with the ideal gas law or by the predictions of Boyle's law discussed for you later in this chapter.

COMBINED GAS LAW

The ideal gas law deals with a single set of conditions. Most of the time, this is not the most convenient way to handle changes in conditions; conditions do change in the real world.

We can look at a change in conditions by handling the original information as if we were dealing with the ideal gas law. We label the original data as the first set of conditions. The second set of conditions, the change, may be dealt with as the ideal gas law labeled to indicate the second set of conditions. The result would look like the following two equations:

1st conditions   **2nd conditions**

$$P_1 V_1 = n_1 R_1 T_1 \quad \text{and} \quad P_2 V_2 = n_2 R_2 T_2$$

The amount of the gas is n and does not change. So $n_1 = n_2$. The value of R is a constant (the gas constant) and constants are called constants because they do not change; $R_1 = R_2$. The way to make some sense out of this is to rearrange the two ideal gas statements so the nR terms are isolated on one side of the equations; divide both sides of the equations by the T that applies.

1st conditions   **2nd conditions**

$$\frac{P_1 V_1}{T_1} = n_1 R_1 \quad \text{and} \quad n_2 R_2 = \frac{P_2 V_2}{T_2}$$

These two equations are equal to each other because the two nR's are equal to each other. Then the equation for changes in conditions is as below.

$$\frac{P_1 V_1}{T_1} = \frac{P_2 V_2}{T_2} \quad \mathbf{3}$$

This equation is called the combined gas law and is a combination of the laws that handle changes in any two factors of volume, pressure, and/or temperature, as long as the third factor is held constant. Boyle's law, Charles' law, and Avogadro's law are a few of the laws that can be predicted by the combined gas law.

BOYLE'S LAW

Gases can be compressed. Boyle's law describes the relationship between volume and pressure: *At a constant temperature, a fixed mass of gas occupies a volume inversely proportional to the pressure exerted upon it.* The reason that "a fixed weight" is included in the statement is that there is no violation of the law of conservation of mass by the simple manipulation of volume and pressure—molecules of the gas are neither gained nor lost.

Keeping temperature as a constant (unchanging value), the combined gas law equation may be manipulated.

$$\frac{P_1 V_1}{T_1} = \frac{P_2 V_2}{T_2}$$

Since the temperature is constant, temperature will cancel out of the equation. First, let's cross multiply.

$$\frac{P_1 V_1}{T_1} \bowtie \frac{P_2 V_2}{T_2}$$

$$P_1 V_1 T_2 = P_2 V_2 T_1$$

If we divide both sides of the equation by T_2, we end up with this equation.

$$\frac{P_1 V_1 T_2}{T_2} = \frac{P_2 V_2 T_1}{T_2}$$

The T_2's on the left of the equation will cancel. Since $T_1 = T_2$, they will cancel on the right.

$$\frac{P_1 V_1 \cancel{T_2}}{\cancel{T_2}} = \frac{P_2 V_2 \cancel{T_1}}{\cancel{T_2}}$$

We have the mathematical expression of Boyle's law.

$$P_1 V_1 = P_2 V_2$$

For the moment, let's assume that the final volume is required, but we know everything else.. Rearranging to solve for the volume (V_2) produces the result.

$$V_2 = \frac{P_1 V_1}{P_2}$$

SHORTCUT: In the preceding explanation, the two T's are on opposite sides of the equal sign.

$$\frac{P_1 V_1}{\cancel{T_1}} = \frac{P_2 V_2}{\cancel{T_2}}$$

T_1 and T_2 are in two fractions equal to each other and were involved in division. Whenever two fractions are equal to each other and the two like terms are either involved in division or multiplication (not addition or subtraction), they will cancel.

Example 7.3: What is the volume of 1 liter of gas as measured under room conditions (25°C and 760 torr) if the pressure is increased to 800 torr?

Solution: Working this problem requires that you recognize two sets of conditions. If there were only one set of conditions, we would use the ideal gas law.

The clue to our dealing with two sets of conditions is the statement that the pressure changes. One really neat trick in keeping everything straight is to write the symbols that we have to work with directly in the problem statement. Look at this:

$$
\begin{array}{c}
\overset{V_2}{\text{What is the volume}} \text{ of } \overset{V_1}{\text{1 liter}} \text{ of gas as measured under room} \\
\text{conditions (25°C and } \underset{P_1}{760 \text{ torr}} \text{) if the pressure is increased to } \underset{P_2}{800 \text{ torr}}? \\
\underset{\text{one } T}{}
\end{array}
$$

Temperature is given only once, and there is no mention of a temperature change. Since the temperature doesn't change, $T_1 = T_2$.

A close reading of the problem statement will group the values given together as shown with the problem statement. Reading a problem and marking in the variables is a big help when working word problems.

Since it is now clear that we are dealing with two sets of conditions, the ideal gas law won't work and the combined gas law will work.

The next step is to write the formula to be used.

$$\frac{P_1 V_1}{T_1} = \frac{P_2 V_2}{T_2}$$

Using the shortcut, we may cancel out the temperatures because they are the same, $T_1 = T_2$.

$$\frac{P_1 V_1}{\cancel{T_1}} = \frac{P_2 V_2}{\cancel{T_2}}$$

We need to rearrange to solve for V_2, since that is what we don't know. If you use V_2 as the symbol for the unknown instead of x, you can use an $\times$ to indicate multiplication—that just makes life a little easier and will remind you of what the final units are to be. The isolation of V_2 can now be accomplished by dividing both sides of the equation by P_2.

$$\frac{P_1 V_1}{P_2} = \frac{\cancel{P_2} V_2}{\cancel{P_2}}$$

$$V_2 = \frac{P_1 V_1}{P_2}$$

Substituting in by the comments from the problem is the next step; but the original statement of the combined gas law comes from the ideal gas law. The ideal gas law requires the use of atmospheres, not torr, for substitution into P. You can convert to atmospheres *or* make use of the shortcut on the previous page. The conversion from torr to atm requires a division by 760 torr/1 atm in the original equation. Since it has to be done in two places, the units will cancel anyway.

$$\frac{800 \text{ torr}}{760 \text{ }^{\text{torr}}\!/_{\text{atm}}} \quad \text{AND} \quad \frac{760 \text{ torr}}{760 \text{ }^{\text{torr}}\!/_{\text{atm}}}$$

Also, the $^{torr}/_{torr\,/atm}$ cancels because it would be on the right of the solved equation, where P_2 is divided by P_1. The units would cancel because the bottom unit ($^{torr}/_{torr/atm}$) is the same as the top.

So, we can substitute in without going to atm and the setup will work just fine. Notice that we cancel torr, also.

$$V_2 = \frac{760 \text{ torr} \times 1 \text{ L}}{800 \text{ torr}} = 0.95 \text{ L}$$

Regardless of the technique used the pressure units cancel out, giving us only the volume unit of liters (dimensional analysis does work).

Solution Summary:

$$\frac{P_1 V_1}{T_1} = \frac{P_2 V_2}{T_2}$$

$$\frac{P_1 V_1}{P_2} = \frac{P_2 V_2}{P_2}$$

$$V_2 = \frac{P_1 V_1}{P_2}$$

$$V_2 = \frac{760 \text{ torr} \times 1 \text{ L}}{800 \text{ torr}} = 0.95 \text{ L}$$

Suppose we had gotten a unit other than liters in the answer. That would indicate that the problem was not set up or not worked correctly; after all, the only volume units we had been given to work with were liters. Just as a reminder, dimensional analysis is sort of a running check that the problem is set up properly and is being worked properly because the final units have to conform to the predicted answer. In this case, V_2 can't be expressed in either K or torr or any unit other than liters.

CHARLES' LAW

Charles' law is similar to Boyle's law, except the change in conditions is brought about when volume and temperature are allowed to change but pressure is held constant.

An example you may have run across happened to me. I poured about half a liter of bottled water from a warm plastic bottle, capped the bottle tightly, and put it in the refrigerator. When I went to get something out of the fridge, I noticed that the bottle had collapsed. The drop in temperature caused the pressure to drop in the thin, flexible bottle and it responded by collapsing above the water.

Example 7.4: You have 13.5 grams of a particular gas at 760 torr and measured it to occupy 10.7 liters of volume at 25°C. What is the volume if the temperature is allowed to come to 100°C?

Solution: What is interesting about this problem is that the 13.5 g is not needed and the volume can be predicted to increase by simple logic—hotter gases occupy more space than cooler ones. The mathematical solution should present a higher value than 10.7 liters (this "guesstimate" will help in checking the answer).

One approach to setting up two-sets-of-conditions-type problems is to outline the conditions.

Try this chart:

	1st set	**2nd set**	**Comments**
P	760 torr	760 torr	Note: no P change
V	10.7 L	??? L	
T	25 + 273	100 + 273	Note: converted to K

Often, a chart will help you understand what the problem calls for. The next step is to write the equation we are going to use.

$$\frac{P_1 V_1}{T_1} = \frac{P_2 V_2}{T_2}$$

Since there is no change in pressure, the P's cancel. Let's substitute into the equation as is, instead of solving for the unknown value in symbols, and see what happens.

$$\frac{10.7 \text{ L}}{298\text{K}} = \frac{V_2}{373\text{K}}$$

Cross multiplication gives you the following:

$$10.7 \text{ L} \times 373\text{K} = V_2 \times 298\text{K}$$

One simple technique that can be used to move the unknown term to the left is to reverse the equation. Reversing an equation does not change the equation, just the appearance of the equation.

$$V_2 \times 298\text{K} = 10.7 \text{ L} \times 373\text{K}$$

The next step is to remove the 298K from the left by dividing both sides by 298K.

$$\frac{V_2 \times 298\text{K}}{298\text{K}} = \frac{10.7 \text{ L} \times 373\text{K}}{298\text{K}}$$

At this point, notice that the temperature unit, K, canceled out on both sides leaving only liters. Since volume is expected to be in liters, the problem is being worked properly.

$$V_2 = 13.39 \text{ L}$$

The volume did increase, as would be expected with an increase in temperature (and constant pressure).

___

GAY-LUSSAC'S LAW

In this law, the volume doesn't change, but pressure and temperature are variable. This law is used in the choice of materials with which to build storage tanks for gases. For example, an above-ground propane tank outside a home has to be strong enough to take the increase in pressure during hot days; also, the tank can't collapse when the pressure drops on cold days. Either occurrence would lead to a potentially dangerous leak.

We should write the combined gas law and cancel out the variable that is unchanging (opposite sides of the equation doing the same thing).

$$\frac{P_1V_1}{T_1} = \frac{P_2V_2}{T_2}$$

We have the mathematical statement that expresses Gay-Lussac's law.

$$\frac{P_1}{T_1} = \frac{P_2}{T_2}$$

Remember that temperature (T) is in the Kelvin scale.

Example 7.5: If the pressure of a very strong container is raised from 1 atm at room temperature to 5 atm, what will be the final temperature?

Solution: One look at the wording for this problem tells us that $V_1 = V_2$, which cancels, and

$$P_1 = 1 \text{ atm} \quad P_2 = 5 \text{ atm} \quad T_1 = 25°C \ (+273) \quad T_2 = ??$$

As with all of the gas law problems, the temperature has to be calculated in the Kelvin scale. This problem also uses room temperature, which is 25°C. We have to use K—the conversion from °C to K is an *addition*; additions do not cancel.

The next thing to do is to write the formula and substitute in the information.

$$\frac{P_1}{T_1} = \frac{P_2}{T_2}$$

$$\frac{1 \text{ atm}}{298K} \diagup\!\!\!\!\!\times \frac{5 \text{ atm}}{T_2}$$

Cross multiplication will produce the linear form of the equation. Notice that atm cancel—they are on opposite sides.

$$\frac{1 \, \cancel{\text{atm}}}{298K} = \frac{5 \, \cancel{\text{atm}}}{T_2}$$

$$T_2 = 5 \times 298K$$

The answer in K is what we want, so we're in good shape.

$$T_2 = 1490K$$

The answer to Example 7.5, 1490K, sounds ridiculously high. However, the temperature increase assumes that the increase in temperature does not allow for any heat loss in the process. Needless to say, there is a great deal of heat produced when the pressure increases with no volume change. That is why a diesel engine can run without spark plugs—there is enough heat produced by compression of fuel and air to reach the kindling (ignition) point of the mixture. This is also why SCUBA tanks are placed in water when they are being filled with air—the water removes the heat as it is produced.

DALTON'S LAW OF PARTIAL PRESSURES

Our atmosphere, the air, is composed of a mixture of gases. Suppose the pressure of dry air—dry because water content is different from place to place and from time to time—is one atmosphere of pressure. When we look at the composition of dry air, we find that approximately 80% is nitrogen and 20% is oxygen.

Another way of stating the composition of the air is to say that the *partial pressures* of the nitrogen and oxygen add up to one atmosphere—0.8 atm of nitrogen + 0.2 atm of oxygen.

Dalton's law of partial pressure is the simple addition of all of the pressures contributing to the total pressure.

$$P_{total} = P_1 + P_2 + P_3 + \dots \text{ etc}$$

Example 7.6: You have a mixture of gases at 785 torr. The mixture is made up of 61.0% nitrogen, 17.0% oxygen, and 22.0% carbon dioxide by volume. What is the partial pressure of oxygen in the mixture?

Solution: We can use a characteristic of gases: Each of the gases of the mixture is present throughout the entire volume of the container (homogeneous mixture). Since the problem deals with percent by volume, AND *percent* means "parts per hundred," we can assume we have 100 liters of the mixture; which means that the amount of oxygen present is 17 parts out of the total, 100 parts. "Parts out of the total" is another way of saying we are dealing with a fraction. Fractions are parts of the whole and can be multiplied. How about this?

$$\frac{17}{100} \times 785 \text{ torr} = 133.45 \text{ torr partial pressure of oxygen}$$

If we were asked in Example 7.6 to calculate the partial pressures of the nitrogen and carbon dioxide, we would use the same type of setup. Also, the three partial pressures would add up to the total pressure.

Many of the gases that you deal with in the lab are produced by a chemical reaction (or electrolysis) in one container and collected by bubbling through water (or some other liquid that does not dissolve the gas). Of course, the *gas collected* is going to have some water vapor mixed in; we must account for that water vapor in order to figure out how much of the gas is collected (Fig. 7.1).

Figure 7.1
Collection of a gas
from a chemical reaction
through water.

Example 7.7: You collect 150 mL of hydrogen by bubbling the hydrogen through water, which is at 25.0°C. The measured pressure is 760 torr. What is the pressure of dry oxygen gas under these conditions? The vapor pressure of water is 21.3 torr at 25°C.

Solution: The plan to solve this problem is that the water's vapor pressure and the gas's vapor pressure add up to the total pressure.

$$P_{total} = P_{water} + P_{hydrogen}$$

Subtracting P_{water} from each side of the equation will give us the following:

$$P_{total} - P_{water} = P_{hydrogen}$$

Now we may substitute in the problem information.
$$760 \text{ torr} - 21.3 \text{ torr} = P_{\text{hydrogen}}$$
The calculation will give us the partial pressure of the dry hydrogen collected.
$$P_{\text{hydrogen}} = 738.7 \text{ torr}$$

AVOGADRO'S PRINCIPLE

Avogadro suggested that gases of equal volumes at the same temperature and the same pressure have the same number of molecules. If we look at the ideal gas law, as in Example 7.1, we use
$$PV = nRT$$
under standard conditions (1 atm and 0°C converted to the Kelvin scale) to determine the volume of one mole of carbon dioxide. We find the volume to be 22.4 L. The solution to this problem is 22.4 L regardless of the gas we use as long as we choose an ideal gas. We are assuming that all gases are ideal for this principle.

The n in the ideal gas equation is the number of moles of the gas. Suppose we were not given n but had the weight of the gas—a standard laboratory finding. We could do a little reasoning and a little substitution to find the molecular weight. Here's how it would work:

1. n is the number of moles.
2. The number of moles is related to the weight of the sample (g) and the molecular weight (MW).

$$n = \frac{\text{g}}{\text{MW}}$$

3. Substitute into the ideal gas formula, $PV = nRT$.

$$PV = \frac{\text{g}}{\text{MW}} RT \quad OR \quad PV = \frac{\text{g} RT}{\text{MW}}$$

These two expressions are the same equation.

Example 7.8: A particular gas is collected dry at 98°C and 740 torr. One liter of the gas weighs 2.50 g. What is the molecular weight of this gas?

Solution: Since we are dealing with gases and gases do not change by °C, but by Kelvin, we have to convert the temperature. Pressure in torr won't work, either. The ideal gas law requires atmospheres. Let's do them after we substitute into the formula.
$$PV = nRT$$
We don't have n, but we do have a weight and we have to calculate the MW anyway. The n is the same as g/MW; we can substitute into the equation.

$$PV = \frac{\text{g} RT}{\text{MW}}$$

$$\frac{740 \text{ torr}}{760 \text{ torr}/\text{atm}} \times 1 \text{ L} = \frac{2.50 \text{ g} \times 0.0821 \text{ L atm}/\text{mol K}}{\text{MW}} \times (98 + 273)\text{K}$$

Cross multiplication is the next step. (*Note: 740 torr × 1 L on the left of the equation can be expressed as 740 torr L when multiplied.*)

$$740 \text{ torr L} \times \text{MW} = 760 \text{ torr}/\text{atm} \times 2.50 \text{ g} \times 0.0821 \text{ L atm}/\text{mol K} \times (98 + 273)\text{K}$$

We divide through by 740 torr L to isolate MW.

$$MW = \frac{760 \; ^{torr}\!/\!_{atm} \times 2.50 \text{ g} \times 0.0821 \; ^{L \text{ atm}}\!/\!_{mol \text{ K}} \times (98+273)K}{740 \text{ torr L}}$$

Now we can think about canceling out as many of the units as possible to make the equation at least look less busy. The final units are expected to be in grams/mole, the MW units.

$$MW = \frac{760 \; ^{torr}\!/\!_{atm} \times 2.50 \text{ g} \times 0.0821 \; ^{L \text{ atm}}\!/\!_{mol \text{ K}} \times (98+273)K}{740 \text{ torr L}}$$

Just looking at the units tells us that this problem is solved correctly because MW = $^g\!/\!_{mol}$. When we put this through a calculator to arrive at the answer,

$$MW = 78.2 \; ^g\!/\!_{mol}$$

GRAHAM'S LAW

Graham's law deals with the speed of diffusion of gases. Gases will take up whatever volume in which they are enclosed and will do so evenly; it is as if the molecules become spread evenly throughout the volume.

Graham's law says, that for all practical purposes, the larger the molecule (the heavier), the slower it will move, and vice versa. The statement of the law is that the rate of diffusion of a gas is inversely proportional to the square root of its molecular weight. In math terms it would look like this

$$\text{rate of diffusion} = \frac{\text{a constant}}{\sqrt{MW}}$$

The constant is different for different gases. But

$$\frac{\text{rate of diffusion}_{\text{gas 1}}}{\text{rate of diffusion}_{\text{gas 2}}} = \sqrt{\frac{MW_{\text{gas 2}}}{MW_{\text{gas 1}}}}$$

Example 7.9: Which gas will diffuse faster—hydrogen chloride or ammonia—and by how much?

Solution: From the Periodic Table of the Elements, we find that the molecular weight of HCl is 36.4606 g and the molecular weight of NH_3 is 17.03044 g. Since Graham's law predicts that the lighter molecule will diffuse faster, we know that the quicker moving of these two is ammonia.

To calculate how much faster, we need to use the preceding equation; it will give us the ratio of the rates of diffusion.

$$\frac{\text{rate of diffusion}_{NH_3}}{\text{rate of diffusion}_{HCl}} = \sqrt{\frac{MW_{HCl}}{MW_{NH_3}}}$$

All that we have to do is to calculate the molecular weight of hydrogen chloride and ammonia gasses, substitute in, and calculate the answer.

$$\frac{\text{rate of diffusion}_{NH_3}}{\text{rate of diffusion}_{HCl}} = \sqrt{\frac{36.4606}{17.03044}}$$

$$\frac{\text{rate of diffusion}_{NH_3}}{\text{rate of diffusion}_{HCl}} = 1.463184$$

The interpretation is that the rate of diffusion of NH_3, the speed of movement, is 1.46 times faster than that of HCl. We can say that because the 1.46 can be divided by 1 to give us a fraction ($^{1.46}/_1$). Therefore, we would have, speedwise, a 1.46:1 ratio. This ratio means that for every meter that HCl moves, NH_3 will move 1.46 meters.

QUESTIONS AND PROBLEMS

1. What is the major factor involved in the existence of a compound as a gas rather than a liquid or a solid?

2. Many of the gas laws depend on the ideal gas law statements. If you are working a problem using the combined gas law equation, which of the variables do not have to be converted from the English system to the metric system? Why?

3. Which of the variables does not change when using Boyle's law? Charles' law? Gay-Lussac's law?

4. If you have two 5-liter containers, oxygen in one and carbon dioxide in the other, how does the number of molecules compare? (The temperature and pressure are the same in both of the containers.)

5. If you have 10 gases at 10 different pressures in 10 different 1-liter containers and put them into a 10-liter container, what would be the final pressure?

6. Oxygen gas can be separated from a compound like $KClO_3$ by heating in a test tube and collecting in another. What would be the volume of 100 g of O_2 under lab conditions? *(Note: Lab conditions are taken to be 1 atm and 25°C.)*

7. What is the pressure on a 25-L container that is at 0°C and is holding 500 g of argon gas?

8. A 25-liter cylinder of compressed gas is filled at 25°C with 1,500 grams of acetylene (C_2H_2). If the warning statement on the cylinder is "12 atmospheres maximum," will the container take the pressure?

9. What would be the volume of 5 liters of gas under room conditions if the pressure were increased to 950 torr?

10. You have a 25-liter container holding 18 atm at 100°C. What is the final volume if the pressure drops to 10 atm?

11. Suppose you have a 25-L container holding 18 atm at 100°C and increase the pressure to 30 atm by decreasing the volume. What would be the final volume of the container?

12. Two liters of a gas at 30°C and 0.9 atmospheres are decreased to 0.5 liters and the pressure is increased to 3 atmospheres. What is the final temperature in °C?

13. What would be the final temperature of a 15-L cylinder of gas at 750 torr and 18°C if the pressure were increased to 1,550 torr?

14. Calculate the partial pressure of each of the gases of a 35% oxygen and 65% nitrogen mixture if the total pressure is 2.5 atm.

15. What is the partial pressure of each of the gases in a mixture composed of 18% argon, 32% krypton, and 50% xenon under a pressure of 900 torr?

16. If 250 mL of a gas weighs 0.300 g at STP, what is the molecular weight of the gas?

17. An amount of 5.225 g of a gas are collected at 100°C and 2 atm in a 5-L container. Calculate the molecular weight of the gas.

18. Five hundred grams of a gas (MW = 44 g/mol) are placed in an 18-L container at 3 atm. What is the temperature of the container?

19. What is the pressure exerted by 28 g of a gas, molecular weight of 53 $^g/_{mol}$, on the walls of a 50-L container at room temperature?

20. A sample of 100 g of a gaseous compound that is 42.88% carbon and 57.12% oxygen is collected in a 40-liter container at 0°C and 1 atm (STP). What is the molecular formula of the gas?

21. A particular gas is found to diffuse 1.33 times as fast as another. The faster of the two gases is analyzed and has a molecular weight of 17 $^g/_{mol}$. What is the molecular weight of the heavier?

22. Two gases are placed in a tube-shaped container at one end; at the other end is a sensor that is sensitive to both of the gases and can tell them apart. One of the gases gets to the end in 35 minutes; the other gets to the end in 52 minutes. If the slower of the gases has a molecular weight of 235 g/mole, what is the molecular weight of the faster gas?

23. Ammonia, NH_3, will diffuse 1.236 times faster than an unknown hydrocarbon (C_xH_y) that is 92.258% carbon and 7.742% hydrogen. Calculate the empirical formula and the molecular formula.

24. HCl exists as a gas when not dissolved in water or some other solvent. Br_2 is a liquid that will evaporate fairly quickly at room temperature. Which will diffuse at a quicker speed and by how much?

25. A gas diffuses 4.5 times as fast as another. The molecular weight of the faster

gas is 213 $^g/_{mol}$ and it diffuses at the rate of 4.0 $^{cm}/_{min}$. What is the diffusion rate of the slower gas?

26. If you have ever discharged a CO_2 cartridge, you know that it gets very cold very quickly. In fact, the temperature will drop enough to cause frostbite. What information would you have to know to calculate the amount of cooling when the cartridge empties completely and which law (equation) would be best used to do the calculations?

27. Graham's law can be worked out in terms of density instead of molecular weight. Why?

28. You are a chemist. You are asked to design a perfume. Your assignment is to produce a design of a scent that will last for a long time—most of the day. What size molecules are you going to investigate? Why?

29. A gas will diffuse faster at a higher temperature than at a lower temperature. Why?

30. Figure 7.1 shows a picture in which a gas is being collected over water. What is the actual contents of the test tube?

CHEMICAL EQUILIBRIA

• Chemical Equilibria • Equilibrium Constant, K_c •
• Heterogeneous Equilibria • Equilibrium Constant, K_p •
• Questions and Problems •

CHEMICAL EQUILIBRIA

A **STATIC EQUILIBRIUM** is one in which there is no motion. If a 5-kilogram weight is resting on your desk, there is a static equilibrium between the weight and the desk. The weight is being pulled down by the force of gravity. The strength of the desk essentially pushes upward with the same amount of force. If the desk cannot push back with 5 kilograms of force, the desk will collapse. If the desk pushes back with more than 5 kilograms of force, the weight will float up toward the ceiling.

A **DYNAMIC EQUILIBRIUM** exists in the functioning of some chemical reactions. If a reaction is reversible, the *rate of the reaction to the right* is equal to the *rate of reaction to the left* at **EQUILIBRIUM**.

$$A + B \underset{\text{rate left}}{\overset{\text{rate right}}{\rightleftharpoons}} 2C$$

When the reaction is at equilibrium, for every molecule of *A* that reacts with a molecule of *B* to yield two molecules of *C*, two molecules of *C* decompose to yield a molecule of *A* and one of *B*.

Notice that the term *equilibrium* does not mean the same as equality. When we use **EQUALITY**, we mean that the left equals the right. A mathematical equation is the use of equality. If we look at the equation $3W = Z$ and know that $W = 12$, then $Z = 36$. The left side of the equation is 36 and so is the right—an equality.

Equilibrium refers to the rates of reaction, not the amount of material on the left and right of the equation. As a matter of fact, it is very rare for there to be an equal amount of materials on the left and right when a chemical reaction is at equilibrium.

EQUILIBRIUM CONSTANT, K_c

Many chemical reactions are reversible. Equilibrium reactions have a particular type of mathematical relationship, the K_c. The calculation of the K_c is an expression of the relationship of the participants in the reaction. If we have the K_c for a particular chemical reaction, we can calculate the amounts of products and reactants when the reaction has reached equilibrium.

The K_c for a reversible reaction is calculated by multiplying the concentrations of all the products together and dividing by all the reactants' concentrations multiplied.

$$K_c = \frac{[\text{product of the products}]}{[\text{product of the reactants}]}$$

Chemical equilibria are calculated on the basis of **MOLARITY**, moles per liter. The symbol used to indicate molarity is a set of square brackets surrounding the chemical formula; $[H^+]$ means the hydrogen ion concentration in moles/liter.

Example 8.1: Set up the mathematical expression for K_c based on the reaction

$$A + B \rightleftharpoons C + D + E$$

Solution: The K_c is the fraction we get when we divide the products multiplied, [C][D][E], by the reactants multiplied, [A][B].

$$K_c = \frac{[C][D][E]}{[A][B]}$$

Example 8.2: Calculate the K_c for the reaction $PCl_{5(g)} \rightleftharpoons PCl_{3(g)} + Cl_{2(g)}$. When the reaction is at equilibrium in a 6-L sealed reaction vessel, the amounts of reaction participants present are 0.10 mole PCl_5, 0.15 mole PCl_3, and 0.16 mole Cl_2.

Solution: Solving this problem requires you to write the mathematical expression for the calculation of K_c. There are no units associated with K_c.

$$K_c = \frac{[PCl_3][Cl_2]}{[PCl_5]}$$

The square brackets call for you to use units of molarity. The problem statement gives enough information to substitute moles per liter for molarity.

$PCl_5 = 0.10$ mol/6 L $PCl_3 = 0.15$ mol/6 L $Cl_2 = 0.16$ mol/6 L

There are two choices for handling molarity information. One is to calculate the molarity of each gas before substitution into the K_c expression. The other is to substitute directly before calculation. The better choice is to substitute first, then calculate. There is less error introduced by rounding-off at the end than at each step.

If we calculate and then substitute, we round off to some point; this introduces error. The general rule is to round off after the last calculation in a problem.

$$K_c = \frac{\frac{0.15}{6} \times \frac{0.16}{6}}{\frac{0.10}{6}}$$

We do the calculation indicated and round the answer off to two significant digits. *Remember that dividing a fraction by a fraction is solved by inverting the bottom fraction and multiplying the result times the top fraction.*

$$K_c = \frac{\frac{0.024}{6}}{\frac{0.10}{6}} = \frac{0.024}{6} \times \frac{6}{0.10} = 0.040$$

*NOTE: When you are working problems dealing with volume and you see square brackets, you are dealing with **molarity**. Just take the number of moles and divide by the number of liters.*

Example 8.3: The chemical reaction $H_{2(g)} + CO_{2(g)} \rightleftharpoons H_2O_{(g)} + CO_{(g)}$ has a calculated $K_c = 1.60$. One-half mole each of H_2 and CO_2 are placed in a 1-liter reaction vessel and allowed to come to equilibrium with the products. Calculate the concentration of all participants in the reaction at equilibrium.

Solution: As with working many word problems, understanding the problem itself is the headache. Most times, representing the problem information is a great way to put things into perspective. In the case of this problem, a table is a great help.

We are starting with $H_2 + CO_2$ and no other chemicals present. This means that the amount of $H_2 + CO_2$ will be reduced as the H_2O and CO appear in the reaction vessel. The amount of reactant loss and product gain is set by the way the equation is balanced; that is a 1:1:1:1 mole ratio in this reaction.

We set up the table of concentrations *in molarity* present by looking at the **original** concentrations, the **gain/loss** concentrations during the reaction, and the **final** concentrations when the reaction comes to equilibrium.

	H_2 +	CO_2 $\rightleftharpoons$	H_2O +	CO
Original	0.5	0.5	0	0
Gain/Loss	$-y$	$-y$	$+y$	$+y$
Final	$0.5 - y$	$0.5 - y$	y	y

We write the K_c statement for this reaction and substitute in the values from the table.

$$K_c = \frac{[H_2O][CO]}{[H_2][CO_2]}$$

$$1.60 = \frac{(y)(y)}{[0.5 - y][0.5 - y]}$$

$$1.60 = \frac{y^2}{(0.5 - y)^2}$$

The last equation is a square on the right. We simplify this equation by taking the square root of both sides.

$$\sqrt{1.60} = \sqrt{\frac{y^2}{(0.5 - y)^2}}$$

$$1.2649 = \frac{y}{0.5 - y}$$

Cross multiplication gives us a linear equation. Converting 1.2649 to a fraction just requires putting it over 1. Any number over 1 is a fraction.

$$\frac{1.2649}{1} \underset{\times}{=} \frac{y}{0.5 - y}$$

$$y = 1.2649 \times 0.5 - 1.2649 \times y$$

We transpose to collect the y terms, then take the square root and multiply.

$$y + 1.2649 \times y = 1.2649 \times 0.5$$

$$2.2649y = 0.63246$$

We divide both sides by 2.2649 to isolate the y.

$$y = \frac{0.63246}{2.2649} = 0.2792$$

We substitute the molar concentrations into the table we used previously.

	H_2 +	CO_2 $\rightleftharpoons$	H_2O +	CO
	$0.5 - y$	$0.5 - y$	y	y
	$0.5 - 0.28$	$0.5 - 0.28$	$+0.28$	$+0.28$
Final	0.22 M	0.22 M	0.28 M	0.28 M

NOTE: Example 8.3 has a square root function that is carried nearly to the end of the solution. Why? The square root of 1.60 is not a short number. This means that we round off the square root at some point in the problem. If we round off early in the problem, we are introducing error into the problem. The earlier we introduce the error, the greater the error in the answer.

If we do not pick up a calculator until the problem is solved, a minimum amount of error is present in the answer.

So far, we have been dealing with reactions that are balanced with one mole of each of the participants. Reactions that are balanced with other than one mole ratios require special handling. We work with the product of the products and the product of the reactants; a product is the result of multiplication.

Let's look at some reactions that are more complex than the preceding.

Example 8.4: Write the K_c for the reaction.
$$F + 2G \rightleftharpoons 3H$$

Solution: We know how to handle the concentration of F, but 2G and 3H present a problem. Let's rewrite the equation this way:
$$F + G + G \rightleftharpoons H + H + H$$
We can write the 2G as G + G and 3H as H + H + H because multiplication is a fancy form of addition.

All we do is follow the K_c formula to write the mathematical equation. We multiply all of the members of the chemical equation and record them in the K_c formula.

$$K_c = \frac{[H][H][H]}{[F][G][G]}$$

We can clean up the expression for K_c.

$$K_c = \frac{[H]^3}{[F][G]^2}$$

Example 8.5: Supply the K_c formula for the reaction
$$4NH_{3\,(g)} + 5O_{2\,(g)} \rightleftharpoons 4NO_{(g)} + 6H_2O_{(g)}$$

Solution: We recognize that there is more than one mole of each of the participants in the reaction. Furthermore, we are aware that *multiplication is a fancy type of addition* in which the $4NH_3$ is the same as $NH_3 + NH_3 + NH_3 + NH_3$; therefore,

$$K_c = \frac{[NO]^4[H_2O]^6}{[NH_3]^4[O_2]^5}$$

Example 8.6: Gaseous hydrogen and iodine react to produce gaseous hydrogen iodide: $H_{2\,(g)} + I_{2\,(g)} \rightleftharpoons 2HI_{(g)}$. The concentration equilibria are found to be 0.36 M H_2, 0.36 M I_2, and 0.75 M HI. Calculate the K_c for this reaction.

Solution: We write the K_c formula on the basis of the chemical reaction.

$$K_c = \frac{[HI]^2}{[H_2][I_2]}$$

Notice that the HI is squared because there are two of them. We substitute the concentrations into the equation and solve for K_c.

$$K_c = \frac{0.75^2}{(0.36)(0.36)}$$

The right side of the equation is a square relationship. We can take the square root of both sides, but that would leave us with the square root of K_c to take care of later. This is not the best choice. The other method of working the problem is to calculate the value of the fraction on the right side of the equation and end up with the answer equal to K_c; this method is a much easier way to handle the math.

$$K_c = \frac{0.5625}{0.1296} = 4.34$$

Example 8.7: A chemical reaction, $SO_{2(g)} + NO_{2(g)} \rightleftharpoons SO_{3(g)} + NO_{(g)}$, is at equilibrium with the concentrations of 0.35 M SO_2, 0.05 M NO_2, 0.23 M SO_3, and 0.45 M NO. What will be the new concentration of SO_3 if 0.12 moles SO_3 are added to the 1-liter container?

Solution: We have a problem with a change introduced. The first thing to do is to calculate the K_c for the original set of conditions.

$$K_c = \frac{[SO_3][NO]}{[SO_2][NO_2]}$$

$$K_c = \frac{(0.23)(0.45)}{(0.35)(0.05)} = \frac{0.1035}{0.0175} = 5.9143$$

Since we have $K_c = 5.9$, we can handle the change. A table is a good way to show the changes requested by the problem. The introduction of more SO_3 will cause stress. According to **Le CHATELIER'S PRINCIPLE**, the stress is relieved by arriving at a new equilibrium. This reaction shifts to the right, relieving stress and arriving at the new equilibrium. There will be a loss of materials from the right and a gain on the left.

	SO_2	+	NO_2 $\rightleftharpoons$	SO_3	+	NO
Original	0.35		0.05	0.23		0.45
Added				+0.12		
Changes	$+y$		$+y$	$-y$		$-y$
Final	$0.35 + y$		$0.05 + y$	$0.35 - y$		$0.45 - y$

We substitute into the equation for K_c and solve.

$$5.9 = \frac{(0.35 - y)(0.45 - y)}{(0.35 + y)(0.05 + y)}$$

$$5.9 \gtrless \frac{(0.35 - y)(0.45 - y)}{(0.35 + y)(0.05 + y)}$$

$$0.10325 + 2.36y + 5.9y^2 = 0.1575 - 0.80y + y^2$$

$$4.9y^2 - 3.16y + 0.05425 = 0$$

We are left with a quadratic equation. We cannot take the square root of both sides because the left is not a perfect square. We cannot solve this equation by factoring. We have to resort to some other method than just transposing from one side to another. This calls for use of the **QUADRATIC FORMULA**. The preceding equation is already in the form to substitute, $ay^2 + by + c = 0$.

$$y = \frac{-b \pm \sqrt{b^2 - 4ac}}{2a}$$

$$y = \frac{-3.16 \pm \sqrt{3.166^2 - (4 \times 4.9 \times -0.5425)}}{2 \times 4.0}$$

$$y = \frac{-3.16 \pm 3.323989}{9.8}$$

$$y = 0.017, -0.66$$

There are two answers mathematically possible when using the quadratic formula. One of the answers will be nonsense. In the case of this problem, we are to subtract $0.35 - y$. One of the answers is -0.66; this answer makes no sense because negative matter does not work in chemistry. The usable answer is 0.017 molar as the value of y.

The problem requests the final concentration of SO_3. Since $0.35 - y$ is the final concentration of SO_3, we substitute in $(0.35 - 0.017)$ and find that the final concentration of SO_3 is 0.33 M SO_3.

Suppose we have the K_c for a particular reaction. Suppose we look at the same reaction, except that the reaction is reversed. What happens to the K_c? Example 8.8 looks at a reaction we have used in this chapter.

Example 8.8: We are given the reaction $SO_{2(g)} + NO_{2(g)} \rightleftharpoons SO_{3(g)} + NO_{(g)}$. The K_c for the reaction is 5.9. $SO_{3(g)} + NO_{(g)} \rightleftharpoons SO_{2(g)} + NO_{2(g)}$ is the reverse reaction. What is the K_c for the reverse reaction?

Solution: We set up the K expression for the reverse reaction and refer to the K expression in Example 8.7.

$$K_c' = \frac{[SO_2][NO_2]}{[SO_3][NO]}$$

The formula for the two K's are the inverse of each other; therefore, the K_c and K_c' are the inverse of each other.

Since $K_c = 5.9$, then $K_c' = 1/5.9 = 0.17$.

HETEROGENEOUS EQUILIBRIA

A **HETEROGENEOUS EQUILIBRIUM** is one in which there is more than one physical state. In the previous section, all of the products and reactants are in the same physical state; these reactions are **HOMOGENEOUS EQUILIBRIA**. How do we handle an equilibrium in which there is more than one physical state in the sealed reaction vessel?

Equilibria in which one or more participants are solids or liquids and the remainder are gases require special handling. The problem is that solids and liquids are not compressible; they have a *constant molarity*. The values for pure liquids and pure solids are not included in the K_c expressions.

Example 8.9: Set up the K_c formula for the decomposition of ammonium chloride, $NH_4Cl_{(s)} \rightleftharpoons NH_{3 (g)} + HCl_{(g)}$.

Solution: The K_c is calculated on the basis of the *product of the products divided by the product of the reactants*, so long as none of the species involved are pure liquids or pure solids. Pure liquids or pure solids are ignored when we consider the K_c equation.

$$K_c = [NH_3][HCl]$$

Example 8.10: The K_c for $C_{(s)} + CO_{2 (g)} \rightleftharpoons 2CO_{(g)}$ is 1.33×10^{-18}. What is the concentration of CO if the CO_2 concentration is 0.95 M?

Solution: We write the K_c equation, substitute in the given values, and solve.

$$K_c = \frac{[CO]^2}{[CO_2]}$$

$$1.33 \times 10^{-18} = \frac{y^2}{0.95}$$

$$y^2 = 0.95 \times 1.33 \times 10^{-18}$$

$$y^2 = 1.2635 \times 10^{-18} \text{ M CO}$$

We take the square root of both sides and arrive at the answer.

$$y = 1.1241 \times 10^{-9}$$

The problem suggests a two-significant-digit answer.

$$y = 1.1 \times 10^{-9} \text{ M CO}$$

EQUILIBRIUM CONSTANT, K_p

Gaseous equilibria may be expressed in pressure units (atmospheres) rather than in concentration units (molarity). The use of pressure, rather than concentration, is a matter of convenience. Depending on the experiment and available equipment, pressure is much easier to determine than concentration. The subsequent calculation of the K_c can be expressed in terms of pressure (P).

The K_p calculation for the chemical reaction resembles the K_c calculation; we still work with *the product of the products divided by the product of the reactants, but in atmospheres*. Let's look at the following chemical reaction and associated K_p.

$$aS + bT \rightleftharpoons cU + dV$$

$$K_p = \frac{(P_U)^c (P_V)^d}{(P_S)^a (P_T)^b}$$

Example 8.11: The reaction $PCl_{5 (g)} \rightleftharpoons PCl_{3 (g)} + Cl_{2(g)}$ is at equilibrium. Analysis provides us with the partial pressures: $PCl_5 = 0.125$ atm, $PCl_3 = 0.40$ atm, and $Cl_2 = 4.0$ atm. Calculate the K_p.

Solution: We write the formula for the calculation of K_p, substitute, and solve.

$$K_p = \frac{(P_{PCl_3})(P_{Cl_2})}{(P_{PCl_5})}$$

$$K_p = \frac{(0.40)(4.0)}{(0.125)} = 12.8$$

Example 8.12: The partial pressure of CO is 0.0022 atm and of CO_2 is 0.117 atm. What is the value of K_p for the reaction $C_{(s)} + CO_{2\,(g)} \rightleftharpoons 2CO_{(g)}$?

Solution: We notice that this reaction is heterogeneous; the carbon is a solid. The carbon is not included in the K_p calculation. We write the K_p formula, substitute in, and solve for the answer.

$$K_p = \frac{(P_{CO})^2}{(P_{CO_2})}$$

$$K_p = \frac{(0.0022)^2}{(0.117)} = 4.14 \times 10^{-5}$$

---

*NOTE: It is extremely tempting to rush through problems. **Don't rush**. Too many errors occur when a problem is rushed. Actually, one error is too many.*

A way of avoiding error when working problems is to be as neat as you can possibly be. Neatness will help you avoid losing notations critical to the correct solution of a problem. Notations, like the superscripts in equilibria problems, are critical to arriving at the correct solution.

QUESTIONS AND PROBLEMS

1. Write the K_c formula for $SO_{2\,(g)} + NO_{2\,(g)} \rightleftharpoons SO_{3\,(g)} + NO_{(g)}$.

2. Write the K_c formula for $2SO_{2\,(g)} + O_{2\,(g)} \rightleftharpoons 2SO_{3\,(g)}$.

3. Write the K_c formula for $4H_{2\,(g)} + CS_{2\,(g)} \rightleftharpoons CH_{4\,(g)} + 2H_2S_{(g)}$.

4. Write the K_c formula for $4NH_{3\,(g)} + 5O_{2\,(g)} \rightleftharpoons 4NO_{(g)} + 6H_2O_{(g)}$.

5. Write the K_c formula for $COCl_{2\,(g)} \rightleftharpoons CO_{(g)} + Cl_{2\,(g)}$.

6. Write the K_c formula for $2NO_{(g)} + O_{2\,(g)} \rightleftharpoons 2NO_{2\,(g)}$.

7. Write the K_c formula for $CO_{(g)} + H_2O_{(g)} \rightleftharpoons CO_{2\,(g)} + H_{2\,(g)}$.

8. Write the K_c formula for $2NO_{2\,(g)} \rightleftharpoons N_2O_{4\,(g)}$.

9. Write the K_c formula for $2H_{2\,(g)} + 3N_{2\,(g)} \rightleftharpoons 2NH_{3\,(g)}$.

10. Write the K_c formula for $2HI_{(g)} \rightleftharpoons H_{2(g)} + I_{2(g)}$.

11. Calculate the K_c for $COCl_{2(g)} \rightleftharpoons CO_{(g)} + Cl_{2(g)}$ if 0.50 mol $COCl_2$, 0.022 mol CO, and 0.022 mol Cl_2 are present at equilibrium in a 1-L sealed reaction vessel.

12. Carbon monoxide and water react in equilibrium with carbon dioxide and hydrogen; all species are gases. Calculate the K_c for this reaction in a 5.75-L container when 2.3 mol CO, 2.3 mol H_2O, 3.7 mol CO_2, and 3.7 mol H_2 are present at equilibrium.

13. The reaction $SO_{2(g)} + NO_{2(g)} \rightleftharpoons SO_{3(g)} + NO_{(g)}$ is at equilibrium in a 25-L container. Our analysis tells us that 0.85 mol SO_2, 1.35 mol NO_2, 0.25 mol SO_3, and 0.27 mol NO are present. Calculate the equilibrium constant.

14. The K_c for a hypothetical reaction, $A \rightleftharpoons B + C$, is 2.75. Equilibrium concentrations of both B and C are 6 mol/L. Calculate the concentration of A at equilibrium.

15. The K_c for $COCl_{2(g)} \rightleftharpoons CO_{(g)} + Cl_{2(g)}$ is 0.575. There are 1.9 mol of $COCl_2$ introduced into a 1-L reaction vessel. Calculate the concentration of CO at equilibrium.

16. Sulfur dioxide and oxygen are in equilibrium with sulfur trioxide by the following chemical reaction: $2SO_{2(g)} + O_{2(g)} \rightleftharpoons 2SO_{3(g)}$. The contents of a sealed 8-L container are 4.00 mol SO_2, 4.00 mol O_2, and 2.75 mol SO_3. Calculate the K_c.

17. Calculate K_c for $2NO_{(g)} \rightleftharpoons N_{2(g)} + O_{2(g)}$ when a 100-L reaction vessel contains 0.55 mol N_2, 0.55 mol O_2, and 0.072 mol NO at equilibrium.

18. Calculate the K_c for the reaction in which the elements are in equilibrium with ammonia. The amounts present in a 3-liter container are 25.00 mol $NH_{3(g)}$, 4.000 mol $N_{2(g)}$, and 6.500 mol $H_{2(g)}$.

19. $3A + 6B \rightleftharpoons C + 2D + 4E$ is performed in a 3-liter container. The amounts of the reactants and products present at equilibrium are 2 mol A, 4.00 mol B, 6.00 mol C, 6.00 mol D, and 7.40 mol E. Calculate the K_c for this hypothetical reaction.

20. Two moles of hydrogen iodide are introduced into a 1-L container. Calculate the concentration of H_2, I_2, and HI when the $K_c = 45.0$.
$$H_{2(g)} + I_{2(g)} \rightleftharpoons 2HI_{(g)}$$

21. Problem 8 requests the K_c for the equation $2NO_{2(g)} \rightleftharpoons N_2O_{4(g)}$. Write the K_c formula.

22. Problem 10 requests the K_c for the equation $2HI_{(g)} \rightleftharpoons H_{2(g)} + I_{2(g)}$. Write the K_c formula.

23. If the K_c for the equation in Problem #21 is 1.625, what is the K_c'?

24. $K_c = 33.33$ for $COCl_{2(g)} \rightleftharpoons CO_{(g)} + Cl_{2(g)}$. Calculate the K_c'.

25. Write the K_c formula for $CaCO_{3(s)} \rightleftharpoons CaO_{(s)} + CO_{2(g)}$.

26. Write the K_c formula for $C_{(s)} + CO_{2(g)} \rightleftharpoons 2CO_{(g)}$.

27. Write the K_c formula for $Br_{2(l)} \rightleftharpoons Br_{2(g)}$.

28. Write the K_c formula for $H_2SO_{4(l)} \rightleftharpoons SO_{3(g)} + H_2O_{(l)}$.

29. A 3-L reaction vessel is filled with 0.22 mol $NH_{3(g)}$ and 0.22 mol $HCl_{(g)}$ and allowed to come to equilibrium with the product, $NH_4Cl_{(s)}$. The concentrations of $NH_{3(g)}$ and $HCl_{(g)}$ at equilibrium are 0.15 mol each. Write the equation for the reaction as described. Calculate the K_c prime.

30. Mercury can be separated from HgO by being exposed to hot hydrogen gas.
$$HgO_{(s)} + H_{2(g)} \rightleftharpoons Hg_{(l)} + H_2O_{(g)}$$
At a high temperature, the K_c is 11.75. At equilibrium, there are 2 mol H_2 gas present in a 4-liter reaction container. What is the number of moles of gaseous water present?

31. Calculate the K_p for $SO_{2(g)} + NO_{2(g)} \rightleftharpoons SO_{3(g)} + NO_{(g)}$ when there are 1.500 atm SO_2, 1.500 atm NO_2, 0.600 atm SO_3, and 0.400 atm NO measured to be present in a 1-liter sealed reaction vessel.

32. $CO_{(g)} + H_2O_{(g)} \rightleftharpoons CO_{2(g)} + H_{2(g)}$ reaches equilibrium pressures of 2.00 atm CO, 1.25 atm H_2O, 0.25 atm CO_2, and 0.25 atm H_2. Calculate the K_p for this reaction.

33. Present at equilibrium are 5.00 atm HI, 1.25 atm H_2 and 1.25 atm I_2 for the reaction $2HI_{(g)} \rightleftharpoons H_{2(g)} + I_{2(g)}$. Calculate the K_p.

34. A 2-liter reaction vessel contains 10.5 g $NO_{2(g)}$ and 35 g $N_2O_{4(g)}$ at equilibrium at a temperature of 320°C. Calculate the K_p for $2NO_{2(g)} \rightleftharpoons N_2O_{4(g)}$.

35. Calculate the K_p for a reaction in which 2.559 grams of solid ammonium chloride are in equilibrium with 1.7 grams of ammonia and 3.6 grams of hydrogen chloride gases in a 7.5-L container at 250°C.

CONCENTRATION OF SOLUTIONS

• Solution Phases •
• Percentage Solutions (wt/wt) • Percentage Solutions (wt/vol) •
• Ratio Solutions • Parts Per Million and Others •
• Molality • Molarity • Normality • Tonicity •
• Questions and Problems •

SOLUTION PHASES

Solutions are made up of two basic **PHASES** (parts). The **SOLUTE** is the material that is dissolved; the **SOLVENT** is the material in which the solute dissolves. The solvent may be a solid, a liquid, or a gas. The solute may be a solid, a liquid, or a gas. If the amounts of the two materials are fairly close to being the same, it is strictly a matter of choice as to which is the solute and which is the solvent.

SOLVENT	SOLUTE	COMMENTS
Nitrogen	Oxygen	Gas in a gas *The atmosphere*
Water	Carbon Dioxide	Gas in a liquid *Carbonated beverages*
Water	Antifreeze	Liquid in a liquid *Automotive coolant*
Water	Glucose	Solid in a liquid *Intravenous nutrition source*
Alcohol	Iodine	Solid in a liquid *Tincture of iodine*
Iron	Carbon	Solid in a solid *Carbon steel, an alloy*

Table 9.1
Phases of various solutions.

When we discuss solutions in most chemistry courses, the solvent is normally water. As a matter of fact, if you are not told what the solvent is, you may assume that the water is solvent. However, a metal dissolved in a metal is an **ALLOY,** and a solution in which an alcohol is the solvent is a **TINCTURE**.

PERCENTAGE SOLUTIONS (wt/wt)

Percent by weight of a solution is normally the percent of the entire solution, unless you are told differently. Percent is calculated by taking the part, dividing by the whole, and multiplying by 100. In other words, we divide the weight of the solute by the weight of the entire solution and multiply by 100 to get percent solution. There are no units associated with percent solution calculations.

Example 9.1: What does the label read in weight/weight percent solution when a solution is produced by dissolving 5 grams of dextrose in 95 grams of water?

Solution: We set up the general formula for percent solution.

$$\text{percent solution} = \frac{\text{grams solute}}{\text{g solute} + \text{g solvent}} \times 100$$

The next step is performed by substituting into the equation and solving.

$$\text{percent solution} = \frac{5 \text{ g dextrose}}{5 \text{ g dextrose} + 95 \text{ g H}_2\text{O}} \times 100$$

$$\text{percent solution} = \frac{5}{100} \times 100 = 5\%$$

A bottle of the solution described in this problem has the label, "5% Dextrose in Water by Weight."

SHORTCUT: Example 9.1 includes the units in the problem setup. Since there are no units associated with percent solution designations, we can leave out units when working problems.

Example 9.2: What is the percent composition of a solution composed of 50 g NaCl and 150 g water using the weight/weight method?

Solution: We start the solution to this problem by writing the general equation.

$$\text{percent solution} = \frac{\text{grams solute}}{\text{g solute} + \text{g solvent}} \times 100$$

We substitute the appropriate values and solve. We can leave out the units in the setup because the answer is in percent.

$$\text{percent solution} = \frac{50}{50 + 150} \times 100$$

$$\text{percent solution} = \frac{50}{200} \times 100 = 25\%$$

Example 9.3: Calculate the weight/weight percent composition of 236 g sodium nitrate in 3.2 L of water.

Solution: We notice that this problem provides us with a volume of water, 3.2 L. The density of water is considered to be 1.00 g/mL unless we are told differently. Therefore, the weight of the water in this problem is assumed to be 3,200 g.

$$\text{percent solution} = \frac{\text{grams solute}}{\text{g solute} + \text{g solvent}} \times 100$$

$$\text{percent solution} = \frac{236}{236 + 3,200} \times 100$$

$$\text{percent solution} = \frac{236}{3,436} \times 100 = 6.88\%$$

Example 9.4: How do we make up 300 grams of a 0.9% NaCl solution by weight?

Solution: If we were asked to make up 100 grams of solution, we would use 0.9 g NaCl and 91.1 g water to give us the total weight of 100 grams. We recognize that the final weight of the solution is 300 grams; therefore, everything needs to be tripled.

We would place 2.7 g NaCl in a container (flask, beaker, whatever) and add 297.3 grams of water while stirring.

$$\text{percent solution} = \frac{2.7}{2.7 + 297.3} \times 100 = 0.9\%$$

There is one very interesting note to keep in mind when dealing with percent solutions. All calculations ignore the molecular weight of the solute; the calculations address the total number of grams solute and solvent. The calculations do not have anything to do with the actual number of molecules of the solute or the solvent.

Let's consider 100 grams of a 3% solution of glucose in water and a 3% solution of iodine in isopropyl alcohol.

A 3% solution of glucose in water contains the same number of grams as a 3% solution of iodine in isopropyl alcohol. Three grams of glucose is a great deal different from three grams of iodine when we consider the numbers of molecules (or the number of molecular weights); the same applies to the 90 grams of the two different solvents.

A 10% glucose ($C_6H_{12}O_6$) solution would deliver a different amount of sugar than a 10% sucrose ($C_{12}H_{22}O_{11}$) solution. Look at the molecular weights.

The preceding look at the composition of percent solutions tells us that each solution must be considered separately from all others when we are concerned about the number of moles (or molecules) contained in the solution.

Example 9.5: Glucose ($C_6H_{12}O_6$) and sucrose ($C_{12}H_{22}O_{11}$) are both sources of energy. Do you get the same nutrition from a 10% solution (wt/wt) of each?

Solution: No, these two solutions are significantly different from each other. Glucose has a lower molecular weight ($\pm 180\ ^g/_{mol}$) than does sucrose ($\pm 342\ ^g/_{mol}$). If we were to consider the use of the same number of grams of each to produce the solutions, the glucose solution would be stronger (more molecules/gram).

PERCENTAGE SOLUTIONS (wt/vol)

This designation of concentration is often used in the health care fields. Have you ever seen a medical program where an IV (intravenous injection) is administered? Those containers are labeled as percent solute. The calculation method used is the weight in grams of the solute relative to a 100-mL volume of the solution. The solvent used is most often water; however, the percentage solution calculations do not consider the identity of the solvent.

Example 9.6 (*Refer to Example 9.1*): What does the label read in weight/volume percent solution when a solution is produced by dissolving 5 g of dextrose in 95 g of solution?

Solution: We do not know the final volume or density of this solution. We must assume the density of 1 g/mL for water; 95 g solution is 95 mL solution.

We know that we have to calculate with reference to 100 mL of solution. However, we are given 95 mL of solution with which to work. Suppose we look at the water as being 95 mL/100 mL. This fraction of the solution is set up in a ratio and proportion with the 5 g dextrose to calculate as if we had 100 mL solution, not 95 mL.

$$\frac{5 \text{ g dextrose}}{y} \asymp \frac{95 \text{ mL solution}}{100 \text{ mL solution}}$$

$$y = \frac{100 \times 5 \text{ g dextrose}}{95}$$

$$y = 5.26316 \text{ g dextrose}$$

This calculation tells us that there would be 5.26316 g dextrose if we were using 100 m L solution. The label is "5.26 percent dextrose (wt/vol)."

Example 9.7 *(Refer to Example 9.2)*: What is the percent composition of a solution composed of 50 g NaCl and 150 grams water using the weight/volume method?

Solution: Using the logic of Example 9.7, the volume of 150 g is 150 mL. We set up the ratio and proportion to determine the amount of NaCl relative to 100 mL H_2O.

$$\frac{50 \text{ g NaCl}}{y} \asymp \frac{150 \text{ mL solution}}{100 \text{ mL solution}}$$

$$y = \frac{50 \times 100}{150} = 33.3 \text{ g NaCl}$$

This solution is a 33.3% NaCl solution (33.3 g NaCl per 100 mL solution).

Example 9.8 *(Refer to Example 9.3)*: Calculate the weight/volume percent composition of 236 g sodium nitrate in 3.20 L of water assuming no change in volume.

Solution: We set up the ratio and proportion to determine the amount of sodium nitrate relative to 100 mL water. We note that 3.20 L is 3,200 mL for the calculation.

$$\frac{236 \text{ g NaNO}_3}{y} \asymp \frac{3,200 \text{ mL water}}{100 \text{ mL water}}$$

$$y = \frac{100 \times 236 \text{ g NaNO}_3}{3,200} = 7.375 \text{ g NaNO}_3$$

Therefore, a solution of 236 g sodium nitrate in 3.20 L solution is 7.375 g sodium nitrate in 100 mL solution (7.38% $NaNO_3$ solution).

Notice that this technique of calculating concentration has the same omission as the previous weight/weight designation of a solution. There is no consideration given to the amount of the solute as measured by the number of molecules in the solution. If we keep in mind that chemistry works on the molecular level, we are aware that knowing the number of molecules in a sample is very important.

RATIO SOLUTIONS

These solutions are produced by mixing the weight of the solute in the solvent to produce the final volume of the solution. A 1:10,000 adrenaline solution is produced by mixing one gram of adrenaline with water and diluting to 10,000 mL. A 1:10,000 alcohol solution is made by diluting one gram of alcohol out to 10,000 mL. Notice that these solutions do not acknowledge that the solvents in these solutions are different.

How much water is required to do these dilutions? There is no way to determine that in advance. Does it matter what the solute is? Certainly, one gram of isopropyl alcohol ($CH_3CHOHCH_3$) contains a different number of molecules than does one gram of ethyl alcohol (C_2H_5OH). A solution in which the solvent is an alcohol is called a **TINCTURE**.

Example 9.9: What is the procedure for making up a 1:1,000 potato starch solution?

Solution: We keep in mind that ratio solutions are the weight of the solute diluted to the given volume of the solution. Therefore, we weigh one gram of potato starch and dissolve in water and add enough water to end up with 1,000 mL of the solution. We would stir the solution throughout the entire process to make certain that the starch is completely dissolved and that the solution is homogeneous.

NOTE: You probably noticed that water was assumed to be the solvent. Water is assumed to be the solvent unless some other solvent is specified, as previously stated.

Example 9.10: Write the instructions for producing the following solutions:
   (a) 5 liters of a 1:1,000 NaCl solution
   (b) 5 liters of a 1:1,000 $C_6H_{12}O_6$ solution.

Solution: (a) We notice that the 1:1,000 solution requested would have a final volume of 1,000 mL (1 L) requiring one gram sodium chloride. Since the volume requested by the problem is 5 L, all we have to do is multiply everything by five and write the instructions. We measure 5 grams of sodium chloride. While stirring, add sufficient water to obtain a final volume of 5 liters.
   (b) The instructions are the same as part a, except we use $C_6H_{12}O_6$ instead of NaCl.

PARTS PER MILLION AND OTHERS

Parts per million (ppm), which can be in milligrams per liter ($^{mg}/_L$) of solution, is often used for dilute concentrations. This expression of concentration is the same as the preceding methods; there is no provision for different solvents. The unit, mg/L, ignores the character of the solute and the solvent. Industrial pollutants, pollen, ozone, and other components of the atmosphere are often expressed in parts per million.

Solutions that are measured in parts per million include salts in water, especially fresh water. Calcium, magnesium, and iron ions cause "hardness" of water. The concentration of these ions in water is often expressed in parts per million. Lead and mercury are reported in parts per million and, sometimes, parts per billion ($^{mg}/_{1,000 L}$).

The concentration of the common salts in marine waters is often referred to in parts per thousand (ppt, g/L). The concentration of the rarer salts is in parts per million or parts per billion.

Example 9.11: Lead is an extremely dangerous environmental pollutant. How much lead is there in a solution found to be 7 ppm?

Solution: Seven parts per million means that there are 7 grams of lead for every million milliliters of solution; this is a very clumsy way to express the concentration. Another more convenient explanation of 7 ppm is 7 mg/L solution.

Example 9.12: A sample of salt water collected near a seaside beach is found to be 35 parts per thousand. What does that mean?

Solution: Parts per thousand means grams per thousand milliliters. The 35 parts per thousand tells us that there are 35 g salt in 1,000 mL (1 L) of the sample.

MOLALITY

Molality (m) is the first of the measures of concentration that *does* pay attention to what the solute is. The formula for molality is moles solute/kilogram solvent.

$$\text{molality} = \frac{\text{moles solute}}{\text{kilograms solvent}}$$

Concentrations expressed in molality are most often used in discussion of the **COLLIGATIVE PROPERTIES** of a solution. These properties are the physical properties of the solvent that change with increasing concentrations of solute: freezing point depression, boiling point elevation, and others discussed in Chapter 11.

There is a problem with molality. There is no way to determine the final volume of the solution being prepared without actually making up the solution. If we mix a solution with a small amount of NaCl and a relatively large amount of water, then we may assume that there is no volume change. However, if we mix any amount of a liquid, like an alcohol, with a specific amount of water, we would expect a volume change. Molarity, the next topic, solves the problem.

Example 9.13: Calculate the molality of 58 grams NaCl dissolved in 1.5 liters of water.

Solution: We assume that the density of water is 1 g/mL; therefore, 1.5 L of water is considered to be 1.5 kg of water. We also notice that 58 g NaCl is approximately one mole of sodium chloride. All we have to do is to write the equation for molarity, substitute into the equation, and solve for the answer.

$$\text{molality} = \frac{\text{moles solute}}{\text{kilograms solvent}}$$

$$m = \frac{1 \text{ mol NaCl}}{1.5 \text{ kg water}} = 0.67$$

The solution is referred to as a 0.67 m NaCl solution (0.67 m NaCl).

Example 9.14: A solution contains 25 grams NaCl dissolved in 3 liters of water. What is the molality of this solution?

Solution: We convert 25 g NaCl to moles NaCl by dividing by the molecular weight, 58 g/mole. We also recognize that 3 L water is 3 kg water.

$$\text{molality} = \frac{\text{moles solute}}{\text{kilograms solvent}} = \frac{\dfrac{\text{g solute}}{\text{g solute/mol solute}}}{\text{kg solvent}}$$

$$m = \frac{\dfrac{25 \text{ g NaCl}}{58 \text{ g NaCl/mol NaCl}}}{3 \text{ kg water}} = 0.14 \frac{\text{mol NaCl}}{\text{kg water}}$$

$$0.14 \frac{\text{mol NaCl}}{\text{kg water}} = 0.14 \text{ m NaCl}$$

MOLARITY

A molar (M) solution deals with concentration in terms of moles solute per liter solution.

$$\text{molarity} = \frac{\text{moles solute}}{\text{liter solution}}$$

The formula for molarity looks very much like the formula for molality; they both have moles solute in the numerator (top) of the fraction. However, where molality has kg solvent in the denominator (bottom) of the fraction, molarity divides by liter solution. This tells us that there is a significant difference in the way these two solutions are produced.

Molal solution components are weighed and mixed with no real idea of the final volume of the solution. Molar solutions are mixed by weighing the solute and diluting to a specified volume. Since the final volume of the molal solutions *is not* predictable and the final volume of molar solutions *is* predictable, the amount of solute in 1 mL of each of the two solutions (e.g.: 1 m and 1 M) may not be the same.

Example 9.15: Calculate the molarity of a solution containing 3 moles glucose in 12 liters of solution.

Solution: This problem is solved by writing the equation for molarity, substituting the problem information, and solving for the answer.

$$\text{molarity} = \frac{\text{moles solute}}{\text{liter solution}}$$

$$\text{molarity} = \frac{3 \text{ moles glucose}}{12 \text{ liter solution}}$$

$$\text{molarity} = 0.25 \text{ M glucose}$$

Example 9.16: How is 5 L of a 3-M NaCl solution produced?

Solution: We recognize that the 3 M NaCl means that the amount of NaCl required is 3 moles for 1 liter of solution. This solution requested is 5 liters in volume; therefore, we need (3×5) 15 moles NaCl and then dilute to 5 liters.

$$\text{molarity} = \frac{\text{moles solute}}{\text{liter solution}}$$

$$3 \text{ M NaCl} \neq \frac{y}{5 \text{ L}}$$

$$y = \frac{3 \text{ mol NaCl}}{\cancel{\text{L}}} \times 5 \cancel{\text{L}}$$

$$y = 15 \text{ mol NaCl}$$

Example 9.17: What mass of hydrogen chloride is required to produce 50.0 mL of a 6.00-M solution?

Solution: A requirement of molarity is the use of moles in the numerator (top) of the fraction. We are asked to respond to the question in terms of mass. We know that we can express moles as g/mole; we substitute g/mole in the denominator (bottom) of the fraction. The problem presents volume in mL; we convert to liters for the denominator of the equation.

$$\text{molarity} = \frac{\text{moles solute}}{\text{liter solution}} = \frac{g \text{ HCl} / g \text{ HCl/mol HCl}}{\text{liter solution}}$$

$$6 \text{ M HCl} \neq \frac{y / 36.4606 \text{ g HCl/mol HCl}}{0.050 \text{ liter solution}}$$

$$6 \text{ M HCl} \times 0.050 \text{ L} \neq \frac{y}{36.4606 \text{ g HCl/mol HCl}}$$

$$y = 6.00 \text{ M HCl} \times 0.050 \text{ L} \times \frac{36.4606 \text{ g HCl}}{\text{mol HCl}}$$

We cannot cancel units the way things stand. We can substitute mol/L in for M. We see that the substitution and cancellation of units leave us with grams, which is the mass we need to solve the problem.

$$y = 6.00 \frac{\cancel{\text{mol HCl}}}{\cancel{\text{L}}} \times 0.050 \cancel{\text{L}} \times \frac{36.4606 \text{ g HCl}}{\cancel{\text{mol HCl}}}$$

$$y = 10.93818 \text{ g HCl} = 10.9 \text{ g HCl}$$

We mix this solution by dissolving 10.9 g HCl in sufficient water to produce 50 mL of solution.

> *NOTE: Molarity is an extremely important concept that reappears throughout the remainder of chemical studies, including equilibrium reactions and electrochemistry.*

NORMALITY

Normality (N) is a variation on molarity. Normality is defined as gram equivalents of solute per liter of solution.

$$normality = \frac{\text{equivalent weights solute}}{\text{liter solution}}$$

The problem, of course, is to define *equivalent weight* (eq wt), sometimes called a *gram equivalent weight*. An equivalent weight's definition depends on the substance under consideration.

When we are working with acids, we consider the hydrogen ion concentration, $[H^+]$. If we are working with bases, we look at the hydroxide ion, $[OH^-]$. After all, we use acids for the hydrogen ions released, and we use bases for the hydroxide ions available for reaction.

If we are dealing with a monoprotic (one hydrogen per molecule) acid, the normality and molarity are of the same numerical value. In the case of a diprotic acid, the normality is twice the molarity. (Refer to Table 9.2.)

	Example	**Molarity**	**Normality**
Monoprotic	HCl	1	1
		0.5	0.5
Diprotic	H_2SO_4	1	2
		0.5	1.0
Triprotic	H_3PO_4	1	3
		0.5	1.5

Table 9.2
Comparison of molarity to normality for acids.

The normality of bases works in the same way, except that we consider the hydroxide ion concentration. Then 0.25 M $Ca(OH)_2$ would be 0.5 *N* $Ca(OH)_2$.

Example 9.18: A reagent bottle has a label that tells us that the contents is 12 M H_2SO_4. What is the normality of the acid?

Solution: Since sulfuric acid contains two hydrogen ions per molecule, the normality is twice the molarity. Therefore, the 12 M H_2SO_4 solution is 24 *N* H_2SO_4.

Example 9.19: How do we mix 2 liters of a 3.00 *N* $HC_2H_3O_2$ solution from the pure acetic acid?

Solution: Since acetic acid is a monoprotic acid, the numerical values of molarity and normality are the same; therefore, three equivalent weights of acetic acid per liter of solution are required. The solution has a final volume of 2 liters; we weigh 6 equivalent weights of acetic acid to mix with sufficient water to achieve the final volume of 2 liters.

The molecular weight of acetic acid is 60.0524 grams. Six equivalent weights of acetic acid is (6 eq wt × 60.0524 g/eq wt) 360.3144 grams of acetic acid. We mix 2 L of 3 *N* $HC_2H_3O_2$ by weighing 360 g $HC_2H_3O_2$ and diluting to 2 L final volume while stirring thoroughly.

Example 9.20: What mass of sulfurous acid is necessary to produce 50 mL of 4 *N*

H_2SO_3? (Assume that pure sulfurous acid is available.)

Solution: We set up this problem from the purely mathematical standpoint by writing the formula for normality.

$$\text{normality} = \frac{\text{equivalent weights } H_2SO_3}{\text{liter solution}}$$

$$4N \ H_2SO_3 = \frac{y}{0.050 \ L}$$

We cross multiply and substitute eq wt/L for N so that we may cancel units. We want the units in the answer to be in mass units.

$$y = \frac{4 \text{ eq wt } H_2SO_3}{L} \times 0.050 \ L$$

$$y = 0.20 \text{ eq wts } H_2SO_4$$

We convert eq wt H_2SO_3 to grams H_2SO_3. We use dimensional analysis to allow the conversion to solve itself. Notice that one equivalent weight of H_2SO_3 is one-half of the molecular weight. That mass is the amount of H_2SO_3 that will release one mole of hydrogen ions, assuming complete ionization.

$$\frac{82.08 \text{ g } H_2SO_3}{2 \text{ eq wt}} \times 0.20 \text{ eq wt} = 8.2 \text{ g } H_2SO_3$$

An important application of the concept of normality is calculating the volume of a specific solution that will just react with a specific volume and concentration of another solution. Let's look at that kind of problem.

Example 9.21: What is the volume of $0.15 \ N \ HNO_3$ that will just neutralize 10 mL of $0.001 \ N \ Ca(OH)_2$ solution?

Solution: The relationship between volume and normality is expressed by a simple equation.

$$V_1 N_1 = V_2 N_2$$

The units of volume are of convenience; the units may be either in milliliters or liters. We substitute into the equation and solve for the answer.

$$V_1 \times 0.15 \ N_1 = 10 \text{ mL} \times 0.001 \ N_2$$

$$V_1 = \frac{10 \text{ mL} \times 0.002 \ N_2}{0.15 \ N_1}$$

$$V_1 = 0.067 \text{ mL } HNO_3$$

TONICITY

The concentration of solutions can be compared using *tonic* and qualifying with a prefix to modify the meaning; *tonic* refers to the concentration of the solute. The terms **ISOTONIC**, **HYPOTONIC**, and **HYPERTONIC** are relative terms; we must be comparing two or more solutions against each other or the use of the terminology is meaningless.

Solutions have two phases—solvent and solute. When we describe solutions using *isotonic*, *hypotonic*, and/or *hypertonic*, we are *defining* the solutions directly in terms of the solute, but we are also *implying* the concentrations of the solute. Table 9.3 presents the terminology.

PREFIX	ROOT	DEFINITION
Iso	tonic	Isotonic solutions are composed of the same concentration of solute. *Implied is the same concentration of solvent.*
Hypo	tonic	Hypotonic solutions are of lower concentration of solute than the solutions against which they are compared. *Implied is that the concentration of solvent is higher than the other solution.*
Hyper	tonic	Hypertonic solutions are of higher concentration of solute than the solutions against which they are compared. *Implied is that the concentration of solvent is lower than the other solution.*

Table 9.3
Classification of solutions.

There is a reasonably easy way to remember the meaning of the prefixes. *Hyper* is easy; a super-active child is a *hyper*active child. *Hypertonic* is used to describe the *higher concentration of solute*. *Hypo* is pinned down because injections are given by means of a *hypo*dermic syringe; injections are below the skin. Then a *hypotonic* solution is the one that is *lower (below) in concentration of solute*. The definition of *iso* can be remembered by default as "the same" if you keep the other prefixes firmly in mind.

The importance of this concept, tonicity, is tied into **OSMOSIS** (the tendency of molecules to move from a location of greater concentration to a location of lesser concentration through a *semipermeable* (selective) membrane).

Two solutions of equal concentration of solute, *isotonic solutions*, are involved in a *dynamic equilibrium*. Molecules move at the same rate into and out of a volume. In other words, there is no net (overall) change in concentration (Figure 9.1).

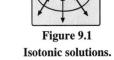

Figure 9.1
Isotonic solutions.

Hypertonic and *hypotonic* are terms that must be used together since they are used to compare two solutions against each other. A balloon is made of a semipermeable membrane filled with a 10% glucose solution and placed in a container of distilled water. The balloon will tend to increase in size as the water moves into the balloon (Fig. 9.2); *water moves from hypotonic to hypertonic.* The balloon's contents have a lower concentration of water than the outside; water will flow through the membrane until the two solutions are isotonic or the membrane bursts.

If the balloon is filled with distilled water and placed in a 10% glucose solution, water moves out the balloon. The balloon will shrink (Fig. 9.3).

Figure 9.2
Hypertonic
balloon.

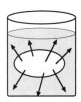

Figure 9.3
Hypotonic
balloon

These concepts are important in storing of living tissues (blood), calculating solution concentrations for injection, and purifying water. Water purification can be performed by *reverse osmosis*, which requires an input of energy.

Example 9.22: You are preparing for a party, and you are going to serve celery with a dip. What is the best way to store the celery to maintain crispness?

Solution: Storing the celery under tap water would keep the celery crisp. The crispness is due to the water content of the celery. Tap water is hypotonic to the celery; water would tend to move into the celery tissues, keeping the crispness. It is also a good idea to store celery in the refrigerator to reduce the celery's rate of metabolism and retard decay.

QUESTIONS AND PROBLEMS

1. A percent solution by weight to weight is produced by mixing 10 grams of table salt, NaCl, with 90 grams of water. What is the percent table salt in this solution?

2. Calculate the % H_2SO_4 by weight in a solution of 175 g H_2SO_4 and 500 g H_2O.

3. What is the percent composition by weight of a solution produced by mixing 1.25 g I_2 and 150 g isopropyl alcohol?

4. How do you produce 200 g of a 25% (wt/wt) sucrose solution?

5. What are the components of 1 kg of a 15% NaCl solution by weight?

6. A weight to volume solution is produced by mixing 25 g glucose and 250 mL water. What is the percent composition of glucose?

7. Calculate the percent HCl by weight/volume in a solution produced from 35 g HCl and 750 g H_2O.

8. What is the percent composition by weight to volume of a solution produced by mixing 250 g tar with 1 L gasoline?

9. How do you produce 200 mL of a 25% (wt/vol) sucrose solution?

10. What are the components of 1 liter of a 15% NaCl solution by weight to volume?

11. What is the amount of $C_{12}H_{22}O_{11}$ in 1 L of a 1:1,000 solution?

12. How much sodium chloride is required to produce 12 L of a 5:10,000 solution?

13. Explain the meaning of 4 ppm mercury in a sample of water from a lake.

14. A 1-liter sample of water contains 28 grams of a salt. How is this solution referred to in terms of parts per thousand?

15. Parts per million is an expression of the number of milligrams of solute that are in 1 liter of solution. What is the meaning of parts per billion?

16. We dissolve 0.9 g NaCl in 100 mL of water. How is this solution labeled in parts per thousand?

17. What is the weight of NaCl in a $12^o/_{oo}$ solution?

18. How do you mix 1 L of a 0.750 molal NaF solution? Assume no volume change.

19. Assuming no change in volume during the production of the solution, calculate the molality of a solution produced by mixing 180 g $C_6H_{12}O_6$ in 2 liters of water.

20. What is the molality of a solution containing 1.2 g sodium sulfate in 50 mL water?

21. A solution is produced by mixing 100 g CH_3OH, methyl alcohol, and 600 g of water. Calculate the molality of this solution.

22. Calculate the molality of a mixture of 50 g octane, C_8H_{18}, in 250 g acetone.

23. Express the molarity of 10 L of a solution that contains 25 g NaCl.

24. A solution has 300 g $C_6H_{12}O_6$ dissolved in sufficient water to produce 37.5 L of solution. Calculate the molarity of this glucose solution.

25. Write the instructions for mixing 5.0 L of 0.020 M Na_2S.

26. Describe the production of 100 L of 12 M H_2SO_4 solution.

27. How many liters of 0.50 M $Mg(NO_3)_2$ solution is produced from 25 g magnesium nitrate?

28. What volume of 1.05 M $HC_2H_3O_2$ can be mixed when 12.5 g acetic acid and sufficient water are provided?

29. You are assigned to mix 7.2 L of synthetic stomach acid. A source on human physiology tells you that the acid is 0.100 M HCl. How do you mix the synthetic somach acid solution?

30. What is the mass of LiI present in 15.0 mL of 0.350 M LiI solution?

31. Calculate the amount of water necessary to produce 0.575 M KNO_3 from 160 g KNO_3.

32. What volume of water is needed to mix a 0.350-M ammonium chloride solution starting with 0.500 g NH_4Cl?

33. What is the normality of a 6-M HCl solution?

34. Calculate the normality of a 3-M sulfuric acid solution.

35. Express the normality of 25 mL of 6 M NH_4OH.

36. Calculate the normality of 12 M NaOH.

37. What is the normality of a 0.0015-M $Ca(OH)_2$ solution?

38. What is the volume of 0.10 N HCl solution that will neutralize 50.0 mL of 0.150 N $Mg(OH)_2$ solution?

39. Calculate the volume of 1.320 N HNO_3 that is necessary to neutralize 40.000 mL of 0.1130 M $Al(OH)_3$.

40. Drinking ocean water is very dangerous. Why?

41. Injecting distilled water directly into a muscle is not an accepted practice. Why?

42. Suppose you were working for a blood bank. You remove red blood cells from whole blood by using a centrifuge and have to store the cells for later use. How would you store those cells?

43. Reverse osmosis can be used to purify ocean water sufficiently of the salts so it can be used as drinking water. What is the process?

• Electrolytes •
• Solutions of Strong Electrolytes • Solutions of Weak Electrolytes •
• Five Percent Rule • Solubility Product, K_{sp} • K_{sp} and The Common Ion Effect •
• Questions and Problems •

ELECTROLYTES

There are many substances that dissolve (the solutes) in water (the solvent). Quite a few of those solutes simply go into solution (dissolve) with no major change to the solute. An example of a solute that is not affected by the dissolution process is table sugar, sucrose. On the other hand, there are solutes that release their component ions when dissolved in a polar solvent, such as water.

An **ELECTROLYTE** is a solute that ionizes when it dissolves. If the solute is an ionic compound, such as NaCl, the ionization process is complete. When sodium chloride dissolves in water, the molecule tends to form the hydrogen ion, H^{+1} and the chloride ion, Cl^{-1}. Those solutes that ionize completely (or nearly completely) are **STRONG ELECTROLYTES**.

*NOTE: The **CATION** is the positive ion; the **ANION** is negative.*

NOTE: If we are not specifically told the name of the solvent, we assume that the solvent is water. Water is a polar solvent; it dissolves polar and ionic solutes rather well.

A **WEAK ELECTROLYTE** is a solute that does not ionize completely when in solution. Acetic acid, $HC_2H_3O_2$, is a weak electrolyte. When dissolved in water, acetic acid will release hydrogen cations, H^{+1}, and acetate anions, $C_2H_3O_2^{-1}$; however, there is a small amount of ionization. Acetic acid, even in concentrated solutions, is a weak electrolyte.

A good rule of thumb to identify a strong electrolyte is that a solute that ionizes to the extent of 0.1 M (one-tenth mole per liter) is a strong electrolyte. Ionization below this level tells us we have a weak electrolyte.

There are, of course, degrees of ionization. Moderately strong electrolytes are those that ionize to a greater extent than 0.1 M but do not completely ionize (100% ionization). This manual refers to strong and weak electrolytes.

SOLUTIONS OF STRONG ELECTROLYTES

Working with solutions of strong electrolytes is straightforward. The ionization process goes to completion. We write the equation for ionization of the solute so we have the mole ratios from which we calculate concentrations. We record the given information from the problem at the appropriate place in the equation. We calculate the concentrations of the ions requested by the problem. If the concentration type is not specified, we assume that the concentration units are in molarity (M = $^{mol\ solute}/_{L\ solution}$). Molarity is the most requested unit in these problems.

Example 10.1: What is the concentration of sodium cations in a solution produced from 5.8 grams sodium chloride dissolved in one liter of water? (Assume no volume change in the production of the solution.)

Solution: This solute, NaCl, ionizes completely. As a matter of fact, sodium chloride is an ionic solid. When sodium chloride dissolves water, which is a polar solvent, the ions are released from the crystalline structure. We write the appropriate equation.

$$NaCl \rightarrow Na^{+1} + Cl^{-1}$$

We put in the information we were given from the problem. We do not have the number of moles of NaCl; we do have the amount of solution, one liter. We convert 5.8 g NaCl into mol NaCl as we write the problem information into the equation. We show what occurs during the reaction. We end up with the concentrations of all species on completion of the reaction.

	NaCl	$\rightarrow$	Na^{+1}	+	Cl^{-1}
Problem Info.	$-\dfrac{\dfrac{58\ g}{58.44247\ ^g/_{mol}}}{L}$		0 M		0 M
Reaction	$-\dfrac{\dfrac{58\ g}{58.44247\ ^g/_{mol}}}{L}$	$+\dfrac{\dfrac{58\ g}{58.44247\ ^g/_{mol}}}{L}$		$+\dfrac{\dfrac{58\ g}{58.44247\ ^g/_{mol}}}{L}$	
Completion	0 M		0.099243 M		0.099243 M

We see that grams cancel in the reaction line and that moles go to the top of the reaction. The resulting units are mol/L, molarity (M). Our calculations tell us that one liter of the solution produced from 5.8 g NaCl ionizes to release 0.099 M sodium cations and 0.099 M chloride anions. The answer to the problem is 0.099 M Na^{+1}.

Example 10.2: What is the concentration of all species present when 2.50 grams of sodium sulfate are dissolved in one liter of water? (Assume no volume change.)

Solution: Let's write the balanced chemical reaction and look at the mole ratio.

$$Na_2SO_4 \rightarrow 2Na^{+1} + SO_4^{-2}$$

The balanced reaction tells us that the amount of sodium cations we have is twice the amount of sodium sulfate, and the amount of sulfate anions is equal to the amount of sodium sulfate. These are important considerations for the calculations that give us the final concentrations. *Equations **must** be correct and balanced.*

We place the information with the equation and work toward the answer.

	Na_2SO_4	$\rightarrow$	$2Na^{+1}$	+	SO_4^{-2}
Prob. Info.	$\dfrac{\dfrac{2.50\ g}{142.043\ ^g/_{mol}}}{L}$		0 M		0 M
Reaction	$-\dfrac{\dfrac{2.50\ g}{142.043\ ^g/_{mol}}}{L}$	$+\dfrac{2\left(\dfrac{2.50\ g}{142.043\ ^g/_{mol}}\right)}{L}$		$+\dfrac{2\left(\dfrac{2.50\ g}{142.043\ ^g/_{mol}}\right)}{L}$	
Completion	0 M		0.0352006 M		0.0176003 M

The answer to this problem is in three significant digits, 0 M sodium sulfate (Na_2SO_4), 0.352 M sodium cations (Na^{+1}), and 0.0176 M sulfate anions (SO_4^{-2}). We notice that the ion ratio is 2:1, as the chemical equation predicts.

NOTE: In the two preceding examples, sodium cations are released. Even though the sodium came from different compounds, the chemical reactivity of the sodium is exactly the same in both solutions. The source of an ion has absolutely nothing to do with the future reactions in which it will engage.

SOLUTIONS OF WEAK ELECTROLYTES

It would be simple if all solutions were of strong electrolytes. Needless to say, that is not anywhere near reality. There are a great many compounds that do not ionize completely in water. These are weak electrolytes in solution.

There are two types of situations that we address. The first is the solute that will dissolve very well in water but does not ionize well at all (acetic acid). The second is the solute that does not dissolve well in water, but the amount that does dissolve is considered to ionize completely. Solutions of the second type are acids, bases, or salts. Salts are considered in the next section, "*Solubility Product, K_{sp}*."

Some weak electrolytes, such as acetic acid, do mix well with water but do not ionize well. We work with an equilibrium between the ions released and the compound. Acetic acid in water is in equilibrium with the hydrogen and acetate ions.

$$HC_2H_3O_{2(aq)} \rightleftharpoons H^{+1}_{(aq)} + C_2H_3O_2^{-1}_{(aq)}$$

We handle the calculations for this equilibrium in the same manner as K_c. The equilibrium constant formula is *the product of the products divided by the product of the reactants*. All species are in units of molarity. Remember that pure species (solids and liquids) are not included in the calculations of equilibrium constants.

$$K_i = \frac{[\text{product of the products}]}{[\text{product of the reactants}]}$$

K_i is used as a general purpose symbol for the ionization of weak substances. Most chemists prefer K_a for designating calculations of the ionization of acids. K_b is associated with the ionization of bases.

$$K_i = \frac{[H^{+1}][C_2H_3O_2^{-1}]}{[HC_2H_3O_2]}$$

The ionization of weak electrolytes is small; the equilibrium constant is small (less than 1) because the formula provides a fraction with a large denominator.

NOTE: Be careful. If the calculated value of K_i for a problem has a value over 1, there is probably a mistake in the calculation for the K_i of a weak electrolyte. A K_i value over 1 indicates a strong electrolyte. Actually, if you were to calculate the K_i for Example 10.1 or 10.2, the value would be infinity (∞) because you divide by zero.

Example 10.3: Calculate the K_i for the hypothetical reaction A $\rightarrow$ B + C. The concentrations at equilibrium are 0.3 M A, 0.02 M B, and 0.02 M C.

Solution: We write the K_i formula, substitute in the concentrations, and calculate.

$$K_i = \frac{[B][C]}{[A]}$$

$$K_i = \frac{(0.02)(0.02)}{0.3} = 1.3 \times 10^{-3}$$

Example 10.4: Calculate the K_i of the hypothetical reaction shown here. The equilibrium concentrations are 0.5 M DE_2, 0.004 M D^{+2}, and 0.006 M E^{-1}.

$$DE_2 \rightleftharpoons D^{+2} + 2E^{-1}$$

Solution: We write the K_i formula for the reaction. We notice that there is a 1:1:2 mole ratio. This can be written as $DE_2 \rightleftharpoons D^{+2} + E^{-1} + E^{-1}$ so we realize that the product of the products contains a squared value, $[E^{-1}]^2$.

$$K_i = \frac{[D^{+2}][E^{-1}]^2}{[DE_2]} = \frac{(0.004)(0.006)^2}{0.5} = 0.000000288 = 3 \times 10^{-7}$$

Example 10.5: The K_a for acetic acid is 1.8×10^{-5}. What is the concentration of all species present at equilibrium if we know we mixed a 0.10 M $HC_2H_3O_2$ solution?

Solution: If we write the equation for the ionization of acetic acid and place the given information, what occurs during the reaction, and what is present at equilibrium?

	$HC_2H_3O_2 \rightleftharpoons$	$H^{+1} +$	$C_2H_3O_2^{-1}$
Problem Info.	0.10 M	0 M	0 M
Reaction	$-y$	$+y$	$+y$
Equilibrium	$0.10 - y$	y	y

We write the K_a formula and substitute in the values from the equilibrium line.

$$K_i = \frac{[H^{+1}][C_2H_3O_2^{-1}]}{[HC_2H_3O_2]}$$

$$1.8 \times 10^{-5} \neq \frac{(y)(y)}{0.10 - y}$$

$$1.8 \times 10^{-6} - 1.8 \times 10^{-5}y = y^2$$

We have an equation with no simple solution. The equation is a quadratic relationship. Since we cannot factor, transpose, or use any other straightforward process for solving equations, we make use of the quadratic equation. We collect all terms on the left so the format for the equation conforms with $ay^2 + by + c = 0$. Then, we substitute into the quadratic formula.

$$y^2 + 1.8 \times 10^{-5}y - 1.8 \times 10^{-6} = 0$$

$$y = \frac{-b \pm \sqrt{b^2 - 4ac}}{2a}$$

$$y = \frac{-1.8 \times 10^{-5} \pm \sqrt{(1.8 \times 10^{-5})^2 - 4 \times 1 \times -1.8 \times 10^{-6}}}{2 \times 1}$$

$$y = \frac{-1.8 \times 10^{-5} \pm 2.68334 \times 10^{-3}}{2 \times 1} = 1.3 \times 10^{-3}, \cancel{-1.4 \times 10^{-3}}$$

One of the answers is not a possible answer. We look to the preceding equilibrium concentrations and realize that the negative answer will not work. It will mathematically provide us with more acetic acid than when we started; furthermore, we will have negative amounts of the ions on the right of the equation.

The positive answer is substituted into the preceding equilibrium concentrations. The equilibrium concentrations are 1.3×10^{-3} M H^{+1}, 1.3×10^{-3} M $C_2H_3O_2^{-1}$, and 0.10 M $HC_2H_3O_2$.

Notice that the answer for the acetic acid is 0.09867 M before it is rounded off. The problem calls for two significant digits on the basis of the given information in the problem statement.

FIVE PERCENT RULE

Example 10.5 is solved using the quadratic equation because there is no other apparent way of solving the equation generated to determine K_a. Is there an easier way to solve this problem? Of course there is! There is an exciting mathematical relationship that can be applied to problems where the concentration is much higher than the equilibrium constant.

Example 10.5 evolves into a quadratic relationship because there is a term in the equilibrium line, $0.10 - y$. If we do not have the y included, the equation generated is greatly simplified. The y can be eliminated from consideration if we find that its value is 5% or less than the number from which it is being subtracted, 0.10.

We find the approximate value of y by taking the square root of the equilibrium constant, 1.8×10^{-5}. The square root of K_a is 0.0042426. Five percent of 0.10 is 0.005. Our estimated value of y is less than 5% of 0.10, and y may be eliminated from consideration.

The solution to the problem after we apply the 5% rule is much more direct than applying the quadratic formula.

$$1.8 \times 10^{-5} \neq \frac{(y)(y)}{0.10}$$

$$y^2 = 1.8 \times 10^{-6}$$

$$y = 1.34 \times 10^{-3} = 1.3 \times 10^{-3}$$

The answer from this solution matches the solution from Example 10.5 and, certainly, is a great deal quicker.

SOLUBILITY PRODUCT, K_{sp}

As we have seen in this chapter, the equilibrium constant is calculated by means of the product of the products divided by the product of the reactants. The concentration units are molarity (moles/liter solution). These calculations work when the substances in the equilibrium are in solution (aqueous). What happens if we have an equilibrium in which the reactant is a solid?

Solids have a fixed mole volume. This means that the concentration $(^{moles}/_{liter})$ is a constant. Since equilibrium constant calculations only deal with those species in solution, we omit the solids. K_{sp} is calculated by the product of the products and assumes that the little bit of the solute that does dissolve ionizes completely. The product of the reactants is omitted because the reactant is the solid that does not go into solution to any great extent.

Incidentally, the solubility of solids in water varies with temperature. Solids dissolve better at higher temperatures than at lower temperatures. The tables that describe the solubility of solids are normally written, assuming **LABORATORY CONDITIONS**, one atmosphere and 25°C. If you are not specifically told the conditions, lab conditions are assumed.

Notice that gases are not mentioned in this chapter. The reason is that gases do not have a fixed concentration. The concentration of a gas depends on the temperature, pressure, and volume of the container. Gases are compressible; there is no constant concentration for a gas. Gases conform to their own laws, which are different from those that describe the behavior of solids and pure liquids.

Example 10.6: You attempt to dissolve 1.0 mole of a hypothetical salt, RS, in 1 liter of water at 25°C. You notice that there appears to be no change in the amount of salt you tried to dissolve because you can see it on the bottom of the beaker. You do a very careful analysis of the substances in solution and find that there are 0.0003 moles of R^{+1} and 0.0003 moles of S^{-1}. This salt does not dissolve and ionize well in water. What is the K_{sp} for this hypothetical salt?

Solution: You write the equation for the ionization of RS.
$$RS \rightleftharpoons R^{+1} + S^{-1}$$
You write the K_{sp} formula for this reaction, substitute in data, and solve the equation.
$$K_{sp} = [R^{+1}][S^{-1}]$$
$$K_{sp} = (0.0003)(0.0003) = 9 \times 10^{-8}$$

Example 10.7: Calcium carbonate (limestone, coral, seashells) is considered to be insoluble in water. This is because the solubility is well below 0.1 M. If the K_{sp} for calcium carbonate is 4.7×10^{-9}, what is the solubility of calcium carbonate at saturation?

Solution: As with other chemistry/math problems, we write the chemical equation.
$$CaCO_3 \rightleftharpoons Ca^{+2} + CO_3^{-2}$$
Since we are working with an equilibrium, we write the formula for the equilibrium constant.
$$K_{sp} = [Ca^{+2}][CO_3^{-2}]$$
The K_{sp}, as determined by the equation, tells us that the concentrations of calcium and carbonate ions are the same. Let's use y for these unknown concentrations and solve the problem.
$$4.7 \times 10^{-9} = (y)(y) = y^2$$
Taking the square root of both sides of the equation provides us with the concentration values.
$$y = 6.9 \times 10^{-5}$$
The 6.9×10^{-5} is the concentration (molarity) of calcium cations and of carbonate anions. The mole ratio in the chemical equation tells us that each mole of calcium carbonate that ionizes will release one mole of calcium ions one mole of

carbonate ions. Therefore, the solubility of calcium carbonate is 6.9×10^{-5} molar $CaCO_3$.

There is one problem that occurs when working with solubility product calculations. As long as a square root is required, there is little trouble because your calculator will handle square roots well. What do you do when you are required to take a third root or some other root? Unless there is a special key on your calculator, you can do any root by using the logarithm (log) and antilog (10^x) keys. Consider Example 10.8.

Example 10.8: A hypothetical salt with the formula WR_2 has a K_{sp} of 5.0×10^{-18}. Calculate the solubility of WR_2 and the concentration of the ions present at saturation.

Solution: First we write the equation for the ionization of the salt.
$$WR_2 \rightleftharpoons W^{+2} + 2R^{-1}$$
The next portion of the solution requires the K_{sp} formula for the reaction. We keep in mind that the molar concentration of R^{-1} is twice that of W^{+1} according to the balanced equation.

$$K_{sp} = [W^{+2}][2R^{-1}]^2 \text{ or } K_{sp} = (y)(2y)^2$$

$$5.0 \times 10^{-18} = 4y^3$$

$$y^3 = 1.25 \times 10^{-18}$$

Now that we have a mathematical relationship, we have to solve for the value of y, which calls for the cube root of 1.25×10^{-18}. Let's assume you cannot take a cube root directly on your calculator.

Take the logarithm (log key) of 1.25×10^{-18}. Your screen reads -17.90909. Divide by 3 ($\div 3$) to get -5.9676967. Finally, take the antilog (10^x) and receive the answer to the value of y, 1.0772173×10^{-6}.

The interpretation is that the solubility of WR_2 is 1.1×10^{-6} M. The concentration of W^{+2} is 1.1×10^{-6} M and the concentration of R^{-1} is 2.2×10^{-6} M.

NOTE: It is very easy to lose sight of the goal involved when you calculate a K_{sp} or apply the formula for K_{sp} to a substance that is dissolved. Just keep in mind that the calculations deal with the solute at saturation.

The preceding explanation and examples relating to K_{sp} assume that the ions come from the same salt. Not all K_{sp} calculations are performed with the dissolution of a salt. If you have an existing solution, you can calculate the amount of one ion required to precipitate a concentration of another ion already in solution. Example 10.9 addresses this calculation.

Example 10.9: A solution of 3.0×10^{-14} M chromium III nitrate is mixed. What is the molarity of phosphate that must be exceeded before chromium III phosphate will precipitate? ($K_{sp} = 2.4 \times 10^{-23}$)

Solution: Chromium III nitrate is the original solution mentioned in this problem. All nitrate solutions are soluble in water. The nitrate ion does not enter into this problem because it is a spectator ion.

The chemical equation that does apply to this problem is the disassociation of chromium III phosphate. We write the equation, write the K_{sp} formula, and solve.

$$CrPO_4 \rightleftharpoons Cr^{+3} + PO_4^{-3}$$

$$K_{sp} = [Cr^{+3}][PO_4^{-3}]$$

$$2.4 \times 10^{-23} = (3.0 \times 10^{-14})(y)$$

$$y = 8.0 \times 10^{-10}$$

We need to exceed 8.0×10^{-10} M phosphate in order for any chromium III phosphate to precipitate.

K_{sp} AND THE COMMON ION EFFECT

There is an application of solubility product that was not covered in the previous discussion and examples. We use solubility product calculations to determine the solubility of a substance, most often a salt, in a solution that already contains one of the ions from the substance. **COMMON ION EFFECT** is the concept applied when an ion is provided from different sources. In reality, the common ion effect is an application of Le Chatelier's principle because the additional ion concentration causes a shift in the equilibrium.

Example 10.10: What is the solubility of barium sulfate in a 0.0045-M potassium sulfate solution; $K_{sp} = 1.1 \times 10^{-10}$?

Solution: The K_{sp} for potassium sulfate is not mentioned because Group IA compounds are soluble in water.

We begin the solution to this problem by writing the equation for the disassociation of barium sulfate. We recognize that we are starting with some sulfate ion present. We also recognize that the amount of barium ion concentration tells us the solubility of barium sulfate. We indicate the concentrations, as given in the problem.

$$BaSO_4 \rightleftharpoons Ba^{+2} + SO_4^{-2}$$

	Ba^{+2}	SO_4^{-2}
Original Solution	0	4.5×10^{-3}
During the Reaction	$+y$	$+y$
Equilibrium Concentrations	y	$4.5 \times 10^{-3} + y$

If we follow through with the K formula and substitute in the equilibrium concentration values, we will have a quadratic relationship. It will be a great deal more convenient if we can apply the 5% rule. An approximation of y is the square root of K_{sp}. The square root of K_{sp} is 1×10^{-5} and 5% of the sulfate ion concentration is 2.25×10^{-4}; therefore, y is not significant in relation to the sulfate ion concentration.

The K_{sp} calculation is based on y for $[Ba^{+2}]$ and 4.5×10^{-3} for $[SO_4^{-2}]$.

$$K_{sp} = [Ba^{+2}][SO_4^{-2}]$$
$$1.1 \times 10^{-10} = (y)(4.5 \times 10^{-3})$$
$$y = 2.4 \times 10^{-8}$$

This calculation tells us that the solubility of barium sulfate is 2.4×10^{-8} M.

QUESTIONS AND PROBLEMS

1. Calculate the concentration of Cl^{-1} ions in 1 L of a solution produced from 16.3 g KCl.

2. What is the $[OH^-]$ in a solution composed of 25 g NaOH diluted to 1 liter of volume?

3. Calculate the number of moles of HCl in 32 mL of 6 M HCl.

4. How many moles of KOH are in 500 mL of a 2-M solution of KOH?

5. How many moles of H^{+1} are there in 3 L of a solution that contains 175 g pure H_2SO_4?

6. Express the hydroxide ion concentration in 75 mL of a solution containing 0.575 g LiOH.

7. Calculate the concentration of all species present when 0.10 M HOCN is at equilibrium with the disassociation products; $K_a = 3.5 \times 10^{-4}$.

8. What is the molarity of F^{-1} in a 0.01-M HF solution; $K_a = 7.2 \times 10^{-4}$?

9. Calculate $[H^{+1}]$ in 25 mL of 0.010 M formic acid. The K_a is 1.8×10^{-4}.

10. Calculate the $[PO_4^{-3}]$ when 0.64 M HPO_4^{-2} disassociates. The K_a is 3.6×10^{-13}.

11. A saturated solution of barium sulfate is analyzed and found to contain a concentration of 3.317×10^{-3} M Ba^{+2}. Calculate the K_{sp} for barium sulfate.

12. Calculate the K_{sp} for the saturated solution of MnS that contains 7.1×10^{-8} M manganese ions.

13. A solution of $Bi(OH)_3$ contains 5.9×10^{-11} M Bi^{+3}. Calculate the K_{sp}.

14. The solubility of Fe_2S_3 is 1.1×10^{-18} moles per liter. Calculate the K_{sp} for Fe_2S_3.

15. The chances are that your drinking water is chlorinated to kill disease organisms. Tap water that has been stored for a time can lose chlorine, which evolves out of solution slowly. The water can be tested for the presence of the chloride ion by adding a soluble silver compound. Silver nitrate is normally used for the test. The result is the production of silver chloride, a heavy white precipitate. This test is very sensitive because the K_{sp} for silver chloride is very low, approximately 2.8×10^{-10}. What is the solubility of silver chloride in solution at saturation?

16. If the K_{sp} for lead II sulfide is 1.0×10^{-29}, what is the solubility of PbS in water?

17. What is the concentration of lead ions in solution from lead II sulfate. The K_{sp} is 1.8×10^{-8}.

18. The K_{sp} for $Mg(OH)_2$ is 1.8×10^{-11}. Calculate the solubility of $Mg(OH)_2$.

19. What is the solubility of cadmium hydroxide, $Cd(OH)_2$; $K_{sp} = 1.2 \times 10^{-14}$?

20. Calculate the solubility of $Ni(CN)_2$ in water under lab conditions. The solubility product for nickel II cyanide is 3.0×10^{-23}.

21. What is the molar concentration of iodide ions at saturation for a solution produced from an excess of Hg_2I_2; $K_{sp} = 4.5 \times 10^{-29}$?

22. How many moles of strontium phosphate, $Sr_3(PO_4)_2$, under lab conditions will dissolve in water; $K_{sp} = 1.0 \times 10^{-31}$?

23. What is the solubility of $CdCO_3$ ($K_{sp} = 2.5 \times 10^{-14}$) in 0.0005 M K_2CO_3?

24. CuS is added to a 1.0×10^{-8} M $Cu(NO_3)_2$ solution. Beyond what molar concentration of CuS will precipitation occur? The K_{sp} for CuS is 8.7×10^{-36}.

WATER, ACIDS, BASES, AND BUFFERS

• Water, K_w • What Are pH and pOH? • Weak Acids and Weak Bases •
• Buffer Systems • Henderson-Hasselbalch Equation •
• Questions and Problems •

WATER, K_w

Water is the most common solvent on this planet and water is *the* solvent in biological systems. It is extremely common as a solvent in industrial processes. Water has been extremely well studied because it is so commonly used as a solvent, it is easy to obtain in a pure form, and water is the most important solvent found in living things. Careful studies of water have told us that it is polar and tends to ionize into hydrogen cations and hydroxide anions. It is also a weak electrolyte.

The ionization of a pure solvent, such as water, is called **AUTOIONIZATION**. The autoionization of water can be expressed in terms of an equation.

$$H_2O_{(l)} \rightleftharpoons H^{+1}_{(aq)} + OH^{-1}_{(aq)}$$

A more convenient method of expressing the preceding equation is to include the production of the **HYDRONIUM** ion, H_3O^{+1}.

$$H_2O_{(l)} + H_2O_{(l)} \rightleftharpoons H_3O^{+1}_{(aq)} + OH^{-1}_{(aq)}$$

The reason for using the hydronium ion instead of the hydrogen ion becomes clear in the section that addresses acids and pH later in this chapter.

We can write an equilibrium constant formula for the autoionization of water.

$$K_w = [H_3O^{+1}][OH^{-1}] \quad \text{or} \quad K_w = [H^{+1}][OH^{-1}]$$

Both of these equations say the same thing. We use whichever equation we need depending on the nature of the problem we are working.

Notice that we do not include the concentration of $H_2O_{(l)}$. Pure liquids have a constant concentration and are left out of the K_w formula, as was discussed in previous chapters. K_w is the ion product for the ionization of water in the same way that K_{sp} is the solubility product we discussed in Chapter 10.

The ionization of water is extremely weak and varies with temperature. Pure water releases 1×10^{-7} moles per liter of hydrogen (hydronium) ions and 1×10^{-7} hydroxide ions at 25°C. Ionization is slightly greater at higher temperatures and the ionization of water is slightly less at lower temperatures. $K_w = [H^{+1}][OH^{-1}]$ is the calculation formula at all temperatures.

> NOTE: *Solutions problems in this manual, as in most texts, are written under laboratory conditions (25°C and one atm of pressure).*

The calculation of K_w is performed by substituting into the K_w formula the concentrations of the ions produced. We assume lab conditions.

$$K_w = [H_3O^{+1}][OH^{-1}]$$

$$K_w = (1.0 \times 10^{-7})(1.0 \times 10^{-7})$$

$$K_w = 1.0 \times 10^{-14}$$

When we are working problems that contain strong acids or bases, the autoionization of water does not supply a significant amount of hydronium or hydroxide ions.

WHAT ARE pH AND pOH?

The formal definition of pH is the negative log of the hydrogen ion concentration. The mathematical expression for this definition is

$$pH = -\log [H^{+1}]$$

Let's take a look at the definition and formula. Notice that pH is the negative log of the hydrogen ion concentration. We look at the negative log as the logarithm multiplied by the negative one. This is not the only place in chemistry where we run across the negative log of something. We will look at the pK later in this chapter.

The definition of pH refers to the hydrogen ion concentration, and the mathematical formula tells us the units are in molarity. The square brackets are the signal that we are working with molarity.

Therefore, pH calculations are taking the log of the hydrogen ion concentration in moles per liter (molarity) and multiplying by negative one. The pH scale for weak solutions runs from zero to fourteen. The lower the numerical value, the higher the hydrogen ion concentration and, hence, the stronger the acid solution.

Example 11.1: Which is a more acidic solution, a solution with a pH of 8 or a solution with a pH of 1?

Solution: The solution with a pH of 8 has a hydrogen concentration of 1.0×10^{-8} M H^{+1} (0.00000001 M H^{+1}).

$$pH = -\log [H^{+1}]$$

$$8 = -1 \times \log [H^{+1}]$$

Divide both sides by –1.

$$-8 = \log [H^{+1}]$$

Take the antilog of both sides.

$$\text{Antilog} (-8) = \text{antilog} (\log [H^{+1}])$$

$$[H^{+1}] = 1.0 \times 10^{-8}$$

We use the same mathematical logic and solution to arrive at the hydrogen ion concentration of the solution with a pH of 1. The solution of pH 1 is 1×10^{-1} M H^{+1}.

A pH of 1 indicates a much stronger acid solution than a pH of 8.

---

pOH is the sister concept to pH. The calculation for pOH is the negative log of the hydroxide ion concentration; this calculation is the same as the pH calculation. As with pH, the lower the numerical value, the higher the hydroxide ion concentration.

$$pOH = -\log [OH^{-1}]$$

Example 11.2: Calculate the pOH for a solution that contains 1.0×10^{-6} M OH.

Solution: We write the formula for the calculation of pOH, substitute into the equation, and solve the equation.

$$pOH = -\log [OH^{-1}]$$

$$pOH = -\log 1.0 \times 10^{-6} = 6$$

K_w, discussed in the previous section, is 1.0×10^{-14}. The hydrogen ion concentration from the autoionization of pure water is 1.0×10^{-7}. The hydroxide ion concentration is 1.0×10^{-7}. The negative log of K_w, 1.0×10^{-14}, is 14; the negative log of $[H^{+1}]$, 1.0×10^{-7}, is 7; and the negative log of $[OH^{-1}]$, 1.0×10^{-7}, is 7. This relationship between pK_w, pH, and pOH can be summarized mathematically.

$$-\log K_w = (-\log [H^{+1}]) + (-\log [OH^{-1}])$$

$$\mathbf{pK_w = pH + pOH}$$

$$14 = 7 + 7$$

An interesting observation about this relationship is that the pH and pOH in weak solutions of acids and bases add up to 14. If the pH increases numerically, the pOH decreases so that the sum of the pH and pOH remains 14.

Example 11.3: Calculate the pOH of a solution containing 0.001 M H^{+1}.

Solution: The first step in the solution of this problem is to calculate the pH.

$$pH = -\log [H^{+1}]$$

$$pH = -\log 0.001 = -\log 1.0 \times 10^{-3}$$

$$pH = 3$$

The next step is to calculate the pOH using the relationship of pK_w to pH and pOH.

$$pK_w = pH + pOH$$

$$14 = 3 + pOH$$

$$pOH = 11$$

NOTE: We can compare the concentrations of hydrogen ions and hydroxide ions in Example 11.3. A solution of pH = 3 (1.0×10^{-3} M H^{+1}) contains very little hydroxide ion (1.0×10^{-11} M OH^{-1}).

WEAK ACIDS AND WEAK BASES

There are not many strong acids. A list of the most commonly used strong acids in lab experiments includes nitric, hydrochloric, and sulfuric acids. Nearly all of the other acids commonly used in chemistry lab experiments are weak acids. The way to tell that you are working with a weak acid is to look up the K_a for the acid. If the value of the K_a is less than 1.0×10^{-1}, the acid is usually considered to be a weak acid.

A weak acid is one that does not ionize well. Acetic acid, $HC_2H_3O_2$, is a component found in vinegar and is a weak acid.

$$HC_2H_3O_{2\,(aq)} + H_2O_{(l)} \rightleftharpoons H_3O^{+1}_{\,(aq)} + C_2H_3O_2^{-1}_{\,(aq)}$$

We can write the equilibrium constant, K_c, for this reaction (the product of the products over the product of the reactants).

$$K_c = \frac{[H_3O^{+1}][C_2H_3O_2^{-1}]}{[HC_2H_3O_2][H_2O]}$$

Of course, water in this reaction is a pure liquid and is not included in the final expression of the equilibrium constant. K_c is a constant. The equation can be algebraically rewritten to place the two constants together.

$$K_c[H_2O] = \frac{[H_3O^{+1}][C_2H_3O_2^{-1}]}{[HC_2H_3O_2]}$$

Chemists define $K_c[H_2O]$ as K_a, the constant for the ionization of weak acids. Note that the format for the chemical equation from which we calculate K_a is written with the acid on the left and the ions on the right.

$$HC_2H_3O_{2\,(aq)} \rightleftharpoons H^{+1}_{\,(aq)} + C_2H_3O_2^{-1}_{\,(aq)}$$

K_a for this equation is written omitting the water. Notice that the hydronium ion, H_3O^{+1}, is presented as the hydrogen ion, H^{+1}.

$$K_a = \frac{[H_3O^{+1}][C_2H_3O_2^{-1}]}{[HC_2H_3O_2]}$$

Example 11.4: Calculate the hydrogen ion concentration in 0.015 M $HC_2H_3O_2$. The K_a for acetic acid is 1.81×10^{-5}.

Solution: The ionization of acetic acid gives us the direction we need to solve the problem. We write the equation and substitute in the concentrations from the problem.

	$HC_2H_3O_{2\,(aq)} \rightleftharpoons$	$H^{+1}_{\,(aq)} +$	$C_2H_3O_2^{-1}_{\,(aq)}$
Original Concentrations	0.015	0	0
During the Reaction	$-y$	$+y$	$+y$
Equilibrium Concentrations	$0.015 - y$	y	y

We write the formula for K_a and substitute the equilibrium concentrations.

$$K_a = \frac{[H_3O^{+1}][C_2H_3O_2^{-1}]}{[HC_2H_3O_2]}$$

$$1.81 \times 10^{-5} = \frac{(y)(y)}{(0.015 - y)}$$

Substitution into the K_a formula indicates that we will have a quadratic relationship when we cross multiply. If we can apply the 5% rule to this problem, we may not need to subtract y from 0.015 M, and we would save some time solving the problem. Five percent of 0.015 is 0.00075. We approximate the value of y with the square root of K_a, which is 0.0043. Since the square root of K_a is larger than 5% of 0.015, we cannot ignore y.

$$y^2 = (1.81 \times 10^{-5})(0.015 - y)$$

$$y^2 = 2.7 \times 10^{-7} - 1.81 \times 10^{-5}y$$

We rearrange the equation so we can insert into the quadratic formula.

$$y^2 + 1.81 \times 10^{-5}y - 2.7 \times 10^{-7} = 0$$

$$y = \frac{-b \pm \sqrt{b^2 - 4ac}}{2a}$$

$$y = \frac{-1.81 \times 10^{-5} \pm \sqrt{(1.81 \times 10^{-5})^2 - 4 \times -2.7 \times 10^{-7}}}{2 \times 1}$$

$$y = \frac{-1.81 \times 10^{-5} \pm 1.04 \times 10^{-3}}{2}$$

We notice that the negative answer will not work. The reason is that the negative value (-1.04×10^{-5}) would be added to another negative value (-1.81×10^{-5}).

$$y = 5.1 \times 10^{-4}$$

The hydrogen ion concentration is y, 5.1×10^{-4} M.

Example 11.5: Calculate the pH of 0.015 M $HC_2H_3O_2$.

Solution: This problem is a natural extension of Example 11.4. Sometimes we are given the concentration of an acid and asked for the pH. We have to do a two–step solution to arrive at the pH in that case. Example 11.4 is the first step; this example is the second step.

We recall that pH = $-\log [H^{+1}]$. Then we substitute in the answer from Example 10.4 and complete the calculation.

$$pH = -\log [H^{+1}]$$

$$pH = 5.1 \times 10^{-4}$$

$$pH = 3.3$$

We have a built–in check on our work. Since we have a pH less than 7.0, we know that we have an acid. $HC_2H_3O_2$, of course, is an acid.

The common strong bases are the Group IA bases. Calcium hydroxide and barium hydroxide, depending on the author, are considered to be either strong bases or moderately strong bases. A weak base is generally considered to be one that does not disassociate more than 1×10^{-1} moles per liter. This book uses examples that are clearly either strong bases or weak bases.

As with the handling of weak acids, the equation from which the K_b is calculated is the disassociation of the base. The compound is written on the left of the equation and the ions are written on the right of the equation. K_b is calculated by the product of the products divided by the product of the reactants. The reactants are omitted from the equation because the reactant is a pure solid.

The math for weak bases is handled in the same manner as the math for weak acids.

Example 11.6: Calculate the cobalt ion and hydroxide ion concentration in a saturated cobalt II hydroxide solution; the K_b is 2.5×10^{-16}.

Solution: This problem's solution is based on the equation for disassociation of the base. We write the chemical equation so we know the mole ratio with which we must work.

$$Co(OH)_{2\,(s)} \rightleftharpoons Co^{+2}_{\,(aq)} + 2OH^{-1}_{\,(aq)}$$

We use the equation for the disassociation to write the K_b formula.

$$K_b = [Co^{+2}][2OH^{-1}]^2$$

We are asked for the hydroxide ion concentration and use a symbol for an unknown value to substitute into the K_b formula. Let's use y, as we have been throughout this manual.

$$2.5 \times 10^{-16} = (y)(2y)^2$$

We solve the formula for the value of y.

$$2.5 \times 10^{-16} = 4y^3$$

$$y^3 = 6.25 \times 10^{-17}$$

$$y = 4.0 \times 10^{-6}$$

Substitution of y gives us the concentrations of both ions present at equilibrium.

$$[Co^{+2}] = 4.0 \times 10^{-6} \text{ M}$$

$$2[OH^{-1}] = 8.0 \times 10^{-6} \text{ M}$$

NOTE: Have you noticed how similar Example 11.6 is to the K_{sp} problems we worked in the previous chapter? Since you can work K_{sp} problems, you can easily work problems including weak bases.

Example 11.7: Calculate the pH of a saturated $Co(OH)_2$ solution.

Solution: In Examples 11.4 and 11.5, we have problems that are related in the sense that they follow a natural mathematical pattern of calculation. This problem follows Example 11.6 in the same way; we are asked to calculate the pH from previously determined information, the hydroxide ion concentration.

We must first determine the pOH of the solution. Then we calculate the pH of the solution.

$$pOH = -\log [OH^{-1}]$$

$$pOH = -\log 8.0 \times 10^{-6}$$

$$pOH = 5.1$$

The calculation of pH from pOH makes use of the relationship discussed in the section entitled *"What Are pH and pOH?"*.

$$pK_w = pH + pOH$$

We can either rearrange to solve for pH or substitute directly into the formula and solve. Let's rearrange the formula, substitute into the formula, and solve for the answer.

$$pH = pK_w - pOH$$

$$pH = 14 - 5.1$$

$$pH = 8.9$$

Is this answer reasonable? The disassociation equation tells us that the solution contains hydroxide ions. The presence of hydroxide ions is expected to provide us with a pH above 7.0 (basic solution). The calculated pH of 8.9 is definitely above the pH we expect (over pH = 7.0). Furthermore, a base with the pH of 8.9 is reasonably close to neutral, which is expected with a weak base.

BUFFER SYSTEMS

An acid buffer system is the solution of a weak acid and the strong salt of that weak acid. A basic buffer system is the solution of a weak base and the strong salt of that weak base. Of course, the question is, "What is a strong salt?"

Strong salts are those that will ionize completely when dissolved in water. The strong salts used to produce acid buffers are generally those of Group IA. For example, if the acid used to produce the buffer is acetic acid, the salts most used are sodium acetate or potassium acetate. Sodium and potassium are Group IA elements.

The salts commonly used to produce basic buffers are nitrates or nitrites. The soluble salts of the **HALOGENS** may also be used. The halogens are the lighter Group VIIA elements.

A buffer has the ability to neutralize hydrogen ions and hydroxide ions that are introduced into solution. There is a limit to how much of these ions can be neutralized; that limit is addressed later in this chapter.

The way in which the neutralization occurs is an exercise in the common ion effect and an application of Le Chatelier's principle. Essentially, the ion is neutralized. The neutralization removes ions from the equilibrium established by mixing of the buffer system. The system responds by shifting the equilibrium to maintain stability as related to the equilibrium constant of the acid in an acid buffer system or the base in a basic buffer system.

Let's look at the pH change of an acid buffer system to see what happens when we add a strong acid or a strong base. Since we have worked with acetic acid, a weak acid, let's use a buffer system that is based on acetic acid and a strong salt of acetic acid. We can use sodium acetate, a Group IA salt of acetic acid. The equation for the disassociation of acetic acid is as follows:

$$HC_2H_3O_2 \rightleftharpoons H^{+1} + C_2H_3O_2^{-1}$$

Suppose we add the strong acid, HCl to the solution. Hydrochloric acid ionizes completely to release hydrogen ions and chloride ions. The hydrogen ion increasees in solution, beyond what was originally there, resulting in stress. The reaction shifts to the left to relieve the stress. This is the application of Le Chatelier's principle. The result is that there is a small change in the hydrogen ion concentration but not like anything that would occur from the addition of the HCl to pure water.

The calculations involved center around the K_a formula.

$$K_a = 1.81 \times 10^{-5} = \frac{[H^{+1}][C_2H_3O_2^{-1}]}{[HC_2H_3O_2]}$$

However, we keep in mind that the acetate ion in solution comes from *both* the acetic acid and the sodium acetate. The common ion effect comes into play when we look at the concentrations of all species present in the equilibrium.

Take a good look at Examples 11.8, 11.9, and 11.10. These three examples are related and make a very good point about the functioning of buffer systems. The three examples give us the opportunity to look at the buffer system (Example 11.8); the pH of a strong acid solution (Example 11.9); and the effect of the addition of the hydrogen ion on the buffer system (Example 11.10).

Example 11.8: One liter of an acid buffer system, $HC_2H_3O_2/NaC_2H_3O_2$ is produced by dissolving 0.01 moles $HC_2H_3O_2$ and 0.01 moles $NaC_2H_3O_2$ and diluting to produce one liter of solution (0.01:0.01 molar ratio). Calculate the pH of the buffer system.

Solution: The calculations for the unadulterated buffer system are based on the equation for the disassociation of both the $HC_2H_3O_2$ and the $NaC_2H_3O_2$. This is the situation in which the acetate ion is present from two sources.

$$HC_2H_3O_2 \rightleftharpoons H^{+1} + C_2H_3O_2^{-1}$$
$$0.01-y \qquad\qquad y \qquad\quad y$$
$$NaC_2H_3O_2 \rightarrow Na^{+1} + C_2H_3O_2^{-1} \quad 0.01 \text{ M}$$

We notice that the sodium acetate ionizes complete and that the sodium ion is a spectator ion. The hydrogen ion concentration at equilibrium is y M, and the acetate ion concentration is $y + 0.01$ M. We substitute these values into the K_a formula and solve for the hydrogen ion concentration at the equilibrium of the system.

$$K_a = \frac{[H_3O^{+1}][C_2H_3O_2^{-1}]}{[HC_2H_3O_2]}$$

$$1.81 \times 10^{-5} = \frac{(y)(y + 0.01)}{(0.01 - y)}$$

The application of the 5% rule would be a distinct convenience in working this problem. The square root of K_a is 4.3×10^{-3}, and 5% of 0.01 is 5×10^{-4}. The estimated value of y is well below the threshold required by the 5% rule. We do not have to subtract y from 0.01, nor do we have to add y to 0.01. The K_a formula becomes much simpler.

$$1.81 \times 10^{-5} = \frac{(y)(0.01)}{(0.01)}$$

$$y = 1.81 \times 10^{-5}$$

The value of y is the hydrogen and acetate ion concentrations. pH is calculated by taking the negative log of the hydrogen ion concentration.

$$pH = -\log [H^{+1}]$$

$$pH = -\log 1.81 \times 10^{-5}$$

$$pH = 4.74$$

Alternate Solution: Let's take another look at the K_a formula with the idea of rearranging it for convenience. Since most of the buffer system problems are probably going to ask for pH calculations, let's solve the K_a formula for $[H^{+1}]$ using the acetic acid equilibrium as a model.

$$K_a = \frac{[H^{+1}][C_2H_3O_2^{-1}]}{[HC_2H_3O_2]}$$

$$K_a[HC_2H_3O_2] = [H^{+1}][C_2H_3O_2^{-1}]$$

$$[H^{+1}] = \frac{K_a[HC_2H_3O_2]}{[C_2H_3O_2^{-1}]} \quad \text{or} \quad [H^{+1}] = K_a \frac{[HC_2H_3O_2]}{[C_2H_3O_2^{-1}]}$$

Let's substitute in the equilibrium concentrations and solve for $[H^{+1}]$.

$$[H^{+1}] = 1.81 \times 10^{-5} \frac{(0.01)}{(0.01)}$$

$$[H^{+1}] = 1.81 \times 10^{-5}$$

This answer is the same as we received from the first solution technique. The value of the pH is the same, pH = 4.74. We can use this solution technique with **MONOPROTIC ACIDS** (acids with one hydrogen).

Example 11.9: Calculate the pH of the solution resulting from mixing 0.001 moles HCl in one liter of pure water.

Solution: The pH resulting from the mixture of 0.001 M HCl in one liter of pure water is obtained from the pH formula.

$$pH = -\log [H^{+1}]$$

$$pH = -\log 0.001$$

$$pH = 3$$

A pH of 3 tells us that this is a rather strong acidic solution.

Example 11.10: Calculate the pH of the buffer system after 0.001 moles of $HCl_{(g)}$ are added to the solution. Assume no change in volume when mixing.

Solution: We write the equation for the equilibrium of acetic acid and the ions released. We use the hydrogen and acetate ion concentrations we calculated for Example 10.8.

$$HC_2H_3O_2 \rightleftharpoons H^{+1} + C_2H_3O_2^{-1}$$
$$0.01 \qquad\qquad 1.81 \times 10^{-5} \quad 0.01$$
$$HCl \qquad\rightarrow\qquad H^{+1} + Cl^{-1} \text{ (spectator ion)}$$
$$\text{ionizes completely} \qquad 0.001$$

Note that the preceding equilibrium concentration is the same numerical value as K_a. We assume that the stress of addition of the hydrogen ions from the HCl causes the reaction to shift completely to the left by the molarity of the added hydrogen ions.

	$HC_2H_3O_2$	$\rightleftharpoons$	H^{+1} +	$C_2H_3O_2^{-1}$
From the Buffer	0.01		K_a	0.01
From the HCl			0.001	
Shift to the Left	+0.001		−0.001	−0.001
New Equilibrium	0.011		K_a	0.009

We substitute the new equilibrium concentrations into the rearranged K_a formula and solve for the hydrogen ion concentration.

$$[H^{+1}] = K_a \frac{[C_2H_3O_2^{-1}]}{[HC_2H_3O_2]}$$

$$[H^{+1}] = 1.81 \times 10^{-5} \frac{(0.11)}{(0.009)}$$

$$[H^{+1}] = 2.21 \times 10^{-5}$$

The pH is determined by taking the log of $[H^{+1}]$ and making it negative.

$$pH = -\log [H^{+1}]$$

$$pH = 4.66$$

Example 10.10, coupled with the pH of the buffer system from Example 10.8, tells us that a buffer system can neutralize an appreciable amount of hydrogen ions. Example 10.9 tells us that the pH of the $HCl_{(g)}$ in one liter of solution is 3.0, which is the pH of a strong acidic solution. However, the addition of that same amount of HCl only modified the buffer system by a 0.08 pH unit difference (4.74 to 4.66). What happens if we add a strong base, such as NaOH, to this buffer system?

Example 11.11: Calculate the pH of the buffer system after 0.001 moles of solid NaOH are added to the solution. Assume no change in volume when mixing.

Solution: This problem is a parallel problem to Example 11.10 and is worked out in the same way.

We write the equation for the disassociation of acetic acid. We use the hydrogen and acetate ion concentrations we calculated for Example 10.8.

$$HC_2H_3O_2 \rightleftharpoons H^{+1} + C_2H_3O_2^{-1}$$
$$0.01 \qquad 1.81 \times 10^{-5} \quad 0.01$$

$$NaOH \rightarrow Na^{+1} + OH^{-1}$$
$$\text{ionizes completely} \quad \text{spectator ion} \qquad 0.001$$

The major difference between this problem and Example 11.10 is that the equation shifts to the right. The shift is due to the low ionization rate of water. Acetic acid molecules ionize to provide hydrogen ions, which neutralize the hydroxide ions from the strong base, sodium hydroxide.

As with Example 11.10, the concentration of hydrogen ions begins with the value of K_a.

$$HC_2H_3O_2 \rightleftharpoons H^{+1} + C_2H_3O_2^{-1}$$

	$HC_2H_3O_2$		H^{+1}	$C_2H_3O_2^{-1}$	
From the Buffer	0.01		K_a	0.01	
From the NaOH				$+ OH^{-1}$ 0.001*	
Shift to the Right	−0.001		+0.001*	+0.001	
New Equilibrium	0.009		K_a	0.011	*tied up in HOH

The hydroxide ion concentration added from the NaOH is removed from the solution by the hydrogen ions released from the disassocation of the acetic acid. This is the reason that there is no OH^{-1} indicated on the new equilibrium line.

We substitute the new equilibrium concentrations into the rearranged K_a formula and solve for the hydrogen ion concentration.

$$[H^{+1}] = K_a \frac{[HC_2H_3O_2]}{[C_2H_3O_2^{-1}]}$$

$$[H^{+1}] = 1.81 \times 10^{-5} \frac{(0.009)}{(0.011)}$$

$$[H^{+1}] = 1.48 \times 10^{-5}$$

The pH is determined by taking the log of $[H^{+1}]$ and making it negative.

$$pH = -\log [H^{+1}]$$

$$pH = 4.83$$

As with Example 11.10, there is a very small change in pH. Just how small a change is due to the addition of the 0.001 M OH^{-1}. The pH of 1 L of a 0.001-M NaOH solution is 11, and the effect of the addition of 0.001 moles NaOH to the buffer system was to make the system only 0.9 pH units higher.

Now that we have an idea of how acid buffer systems work, we need to take a look at the basic buffer systems: systems composed of a weak base and the strong salt of that weak base. Keep in mind that basic buffer systems produce OH^{-1} ions. We cannot calculate pH directly. pOH calculated first, and then the pH is determined.

Example 11.12: One liter of a buffer system, NH_3/NH_4Cl, is mixed. NH_3, 0.01 moles, and 0.01 moles NH_4Cl are dissolved and diluted to produce one liter of solution (0.01:0.01 molar ratio). Calculate the pH of the buffer system.

Solution: NH_3 dissolves well in water and enters into a reaction with the water molecules to produce ammonium and hydroxide ions. Instead of writing water as H_2O, let's write water as HOH. Notice that there is one hydrogen lost by the water to the ammonia molecule and that the hydroxide ion is left in solution.

$$NH_3 + HOH \rightarrow NH_4^{+1} + OH^{-1}$$

Ammonium chloride is a strong salt of the ammonium ion. According to the solubility rules, ammonium compounds dissolve and ionize well in water.

$$NH_4Cl \rightarrow NH_4^{+1} + Cl^{-1}$$

Just as we would expect in a buffer system, there is one ion that is being supplied from two sources. That ion is NH_4^{+1}.

The hard way to work this problem is to convert to pHs as we go. The easier way to work the problem is to calculate the pOH and convert to pH. Let's do this problem the easy way. Furthermore, we can parallel the solution technique used in calculating the pH of the acetic acid buffer system in Example 11.8.

We write the disassociation reactions involved in the buffer system. We record the concentration data from the problem statement.

$$NH_3 + HOH \rightarrow NH_4^{+1} + OH^{-1}$$
$$0.01 \qquad\qquad K_b \qquad K_b$$
$$NH_4Cl \rightarrow NH_4^{+1} \qquad\quad + Cl^{-1}$$
$$\text{ionizes completely} \quad 0.01 \qquad\qquad \text{spectator ion}$$

K_b is 1.8×10^{-5}. We have the K_b and the equilibrium concentrations to enter into the K_b formula. Before we get into working with the number, let's write the K_b formula and rearrange for $[OH^{-1}]$. Then we substitute into the formula and solve for the hydroxide ion concentration of the buffer.

$$K_b = \frac{[NH_4^{+1}][OH^{-1}]}{[NH_3]}$$

$$[OH^{-1}] = K_b \frac{[NH_3]}{[NH_4^{+1}]}$$

$$[OH^{-1}] = 1.81 \times 10^{-5} \frac{(0.01)}{(0.01)}$$

$$[OH^{-1}] = 1.8 \times 10^{-5}$$

This is a buffer system produced from a 1:1 molar ratio of the components. If we make use of Example 11.8's alternate solution, we predict that the hydroxide ion concentration would be the numerical value of K_b. The pOH is determined by taking the log of $[OH^{-1}]$ and making it negative.

$$pOH = -\log [OH^{-1}]$$

$$pOH = 4.7$$

The relationship between pH, pOH, and pK_w is used to calculate the pH.

$$pK_w = pH + pOH$$

$$pH = pK_w - pOH$$

$$pH = 14.0 - 4.7$$

$$pH = 9.3$$

As a check on our answer, we notice that the pH of 9.3 is above neutral. This is in the range of a basic substance and seems to be just fine. Had we received a pH below 7.0, we would question the answer, and we would expect an answer in the basic range.

Example 11.13: Calculate the effect of adding 0.001 moles of $HCl_{(g)}$ to the buffer system from Example 11.12. Assume no change in volume when mixing.

Solution: We write the reaction and place the values from the problem statement.

$$NH_3 + HOH \rightleftharpoons NH_4^{+1} + OH^{-1}$$

From the Buffer	0.01		0.01	K_b
From the HCl				$+ H^{+1}$ 0.001
Shift to the Right	−0.001		+0.001	+0.001
New Equilibrium	0.009		0.011	K_b tied up in H_2O

We continue the solution by substituting into the rearranged K_b formula.

$$[OH^{-1}] = K_b \frac{[NH_3]}{[NH_4^{+1}]}$$

$$[OH^{-1}] = 1.81 \times 10^{-5} \frac{(0.009)}{(0.011)}$$

$$[OH^{-1}] = 1.47 \times 10^{-5}$$

The calculation of the pOH depends upon the OH^{-1} concentration.

$$pOH = -\log [OH^{-1}]$$

$$pOH = -\log 1.47 \times 10^{-5}$$

$$pOH = 4.8$$

The calculation of the pH requires the relationship between K_w, pH, and pOH.

$$K_w = pH + pOH$$

$$pH = K_w - pOH$$

$$pH = 14 - 4.8$$

$$pH = 9.2$$

We saw a small change in pH in Example 11.12 due to the addition of the strong acid to the acid buffer system. We found a small pH change (0.1 pH unit drop) in this problem when a strong acid was added to the basic buffer system.

> *NOTE: Although the calculations for buffer systems appear to indicate that the capacity of a buffer to neutralize ions is unlimited, there is a limit. Buffer systems become exhausted when the pH of the system is ±1.0 pH units from pK_a or pK_b, according to the empirically determined data.*

HENDERSON–HASSELBALCH EQUATION

There is one other factor for consideration; that factor comes from the alternate solution for Example 10.8. Take a look at the formula we used for the acetic acid/sodium acetate buffer system. This is a buffer system for a monoprotic acid.

$$[H^{+1}] = K_a \frac{[HC_2H_3O_2]}{[C_2H_3O_2^{-1}]}$$

If the acid and the anion concentrations are the same in the mixture, the value of the hydrogen ion concentration is K_a. This means that if we mix a buffer system in which the weak acid and the strong salt are of the same concentration, the hydrogen ion concentration is equivalent to K_a. This also means that the pH of the system is pK_a.

There are advantages of mixing acid buffer systems that do not have the pH = pK_a. Buffers are used to keep solutions that are sensitive to pH change stable with respect to pH. As an example of required pH stability, consider a test solution used in a medical laboratory. The bottle of test solution is not full; therefore, there is air above the level of the solution. Carbon dioxide is a component of air that will dissolve well in water and react with water to produce carbonic acid, H_2CO_3. The carbonic acid tends to make a solution lower in pH as the carbonic acid disassociates, releasing hydrogen ions. A buffer system that is inert to the testing solution's chemistry tends to keep the pH very stable.

Even though we do have a lot of pH choices when constructing acid buffer systems (lots of pK_a's), there are situations when we need to produce a buffer system that is not at pH = pK_a. We can take another look at the formula used for the calculations of buffer systems and do one more thing to aid us in the determination of the molar ratios required for buffer systems at a pH other than the pK_a. Let us continue to use the acetic acid/sodium acetate system and generalize to other systems later. We use the equilibrium reaction for the disassociation of acetic acid.

$$HC_2H_3O_2 \rightleftharpoons H^{+1} + C_2H_3O_2^{-1}$$

We write the K_a formula for this reaction.

$$K_a = \frac{[H^{+1}][C_2H_3O_2^{-1}]}{[HC_2H_3O_2]}$$

We rearrange this formula to solve for $[H^{+1}]$.

$$[H^{+1}] = K_a \frac{[HC_2H_3O_2]}{[C_2H_3O_2^{-1}]}$$

We used this relationship earlier in the chapter. Let's go one more step and take the log of both sides. The reason will become clear in a moment.

$$\log [H^{+1}] = \log K_a + \log \frac{[HC_2H_3O_2]}{[C_2H_3O_2^{-1}]}$$

The log of the fraction is added to the log of K_a because the step before shows K_a being multiplied by the fraction. Taking a log of multiplied values results in the logarithmic values being added (division results in subtraction).

We multiply both sides of the equation by -1. The reason for the multiplication is that we may now solve directly for pH.

$$-\log[H^{+1}] = -\log K_a - \log \frac{[HC_2H_3O_2]}{[C_2H_3O_2^{-1}]}$$

$$pH = pK_a - \log \frac{[HC_2H_3O_2]}{[C_2H_3O_2^{-1}]}$$

A simple rearrangement gives us a slightly different presentation.

$$pH = pK_a + \log \frac{[C_2H_3O_2^{-1}]}{[HC_2H_3O_2]}$$

We make use of the Brønsted–Lowry concepts relating to acids and bases to write the preceding equation in a generic form. Recall that the acetate anion is the conjugate base of the acid, acetic acid. We substitute these terms into the equation and state the **HENDERSON–HASSELBALCH EQUATION**. The Henderson–Hasselbalch equation is useful in calculations of buffer systems that are not in a 1:1 molar ratio of acid to strong salt.

$$pH = pK_a + \log\frac{[\text{conjugate base}]}{[\text{acid}]}$$

The Henderson–Hasselbalch equation may be expressed for basic buffers also.

$$pOH = pK_b + \log\frac{[\text{conjugate acid}]}{[\text{base}]}$$

Example 11.14: Calculate the molar ratio for the production of an acetic acid/sodium acetate buffer with a pH of 4.2.

Solution: We know from previous examples that the K_a is 1.81×10^{-5} for acetic acid. The pK_a is 4.74. We substitute directly into the Henderson–Hasselbalch equation to determine the requested ratio.

$$pH = pK_a + \log\frac{[\text{conjugate base}]}{[\text{acid}]}$$

$$4.2 = 4.74 + \log\frac{[\text{conjugate base}]}{[\text{acid}]}$$

$$\log\frac{[\text{conjugate base}]}{[\text{acid}]} = 4.2 - 4.74 = -0.54$$

We take the antilog (10^x) of both sides and arrive at the mole ratio requested.

$$\frac{[\text{conjugate base}]}{[\text{acid}]} = 0.29$$

The answer, 0.29, may not look like a ratio but can be changed to one with ease. We can place any number over 1 to make it into a fraction.

$$\frac{[\text{conjugate base}]}{[\text{acid}]} = \frac{0.29}{1}$$

We now have a ratio that requires 0.29 moles/L of conjugate base (sodium acetate) for each mole/L of acetic acid to produce a buffer system with a pH of 4.2.

Molar Ratios $\quad \begin{array}{ccc} HC_2H_3O_2 & \rightleftharpoons & H^{+1} + C_2H_3O_2^{-2} \\ 1 & \text{to} & 0.29 \end{array}$

Example 11.15: Calculate the molar ratio for the production of an NH_3/NH_4Cl buffer with a pH of 10.1.

Solution: This problem asks us to calculate on the basis of a pH. However, we are required to substitute into a formula that calculates on the basis of pOH, the base version of the Henderson–Hasselbalch equation. We calculate the pOH from the pH and use the pOH to calculate the ratio we need to solve the problem.

$$pOH = pK_w - pH$$

$$pOH = 14 - 10.1 = 3.9$$

We substitute into the Henderson–Hasselbalch equation for bases. We use the K_b for the ammonia solution from previous work in this chapter; the K_b for ammonia is 1.8×10^{-5}.

$$pOH = pK_b + \log\frac{[\text{conjugate acid}]}{[\text{base}]}$$

$$3.9 = 4.74 + \log\frac{[\text{conjugate acid}]}{[\text{base}]}$$

$$\log\frac{[\text{conjugate acid}]}{[\text{base}]} = 3.9 - 4.74 = -0.84$$

We take the antilog of both sides and arrive at the mole ratio requested.

$$\frac{[\text{conjugate acid}]}{[\text{base}]} = \frac{0.14}{1}$$

The ratio that satisfies this problem is 0.14 mole/L of the conjugate acid (NH_4Cl) for each mole/L of the base (NH_3) to produce a buffer system with a pH of 10.1.

NOTE: Buffer systems should not be mixed more than ±1 unit from the pK_a or pK_b. They will not work as predicted or won't work at all.

QUESTIONS AND PROBLEMS

1. Calculate the pH of a 0.001-M HCl solution.

2. What is the pH of a 0.0340-M HNO_3?

3. The pH of a sample of black coffee is 4.00. Calculate the molar concentration of hydrogen ions present in the coffee.

4. The pH of human stomach acid is measured to be 2.27. What is the concentration of hydrogen ions?

5. What is the OH^{-1} concentration of a solution in which the pH is 3.2?

6. Calculate the molar concentration of OH^{-1} in a solution with a pH of 5.25.

7. Express the H^{+1} concentration in a solution with a pOH of 7.90.

8. Find the concentration of hydronium ions in a solution if the pOH is 13.25.

9. Household ammonia solution has a pH of approximately 11.0. What are the concentrations of the hydrogen and hydroxide ions present in this solution?

10. Calculate the concentration of both hydrogen and hydroxide ions in a seawater solution that has a pH found to be 8.13.

11. Rainwater is generally close to neutral. Some of the products that go up a smoke stack dissolve in rainwater and react with it. A sample of rain is collected around a manufacturing plant. The pH of the sample is measured as 6.10. What are the hydrogen and hydroxide ion concentrations in this sample of acid rain?

12. Calculate the hydrogen ion concentration of a 0.55-M acetic acid solution. The K_a for $HC_2H_3O_2$ is 1.81×10^{-5}.

13. Calculate the pH of the acetic acid solution in problem #12.

14. Nitrous acid, HNO_2, is a weak acid with a K_a of 4.5×10^{-4}. Calculate the H^{+1} concentration of a 0.75-M solution of nitrous acid.

15. Calculate the pH of the nitrous acid solution from problem #14.

16. What is the OH^{-1} concentration of a 3.00-M solution of ammonia? The K_b for ammonia is 1.8×10^{-5}.

17. Calculate the pH of the ammonia solution from problem #16.

18. A buffer solution is mixed using 0.010 moles each of HOCN and KOCN. Calculate the pH of the buffer system; the K_a for HOCN is 3.5×10^{-4}.

19. What is the pH of a buffer solution produced from 0.0075 M HOBr and 0.0075 M NaOBr? The K_a for HOBr is 2.5×10^{-9}.

20. In what ratio are the acid and salt mixed to produce a buffer with a pH of 4.0 using acetic acid and sodium acetate? $K_a = 1.81 \times 10^{-5}$.

21. A buffer system with a pH of 6.90 is desired. Hypochlorous acid, HOCl, and potassium hypochlorite, NaOCl, are mixed. The K_a for HOCl is 3.5×10^{-8}. What is the ratio of HOCl/NaOCl required?

22. You are attempting to mix a buffer solution with a pH of 4.50. You find that you only have 0.250 moles $HC_2H_3O_2$. You produce one liter of a 0.250-M acetic acid solution. How many moles of potassium acetate do you need to complete the buffer system?

23. Calculate the hydrogen ion concentration after 0.045 moles of sodium hydroxide have been added to one liter of a buffer solution composed of 0.500 moles acetic acid and 0.500 moles of sodium acetate. Assume that no volume changes occur during the experiment.

24. Two and a half liters of a buffer system are produced from the one mole each of HCN and LiCN. The K_a for HCN is 4.0×10^{-10}. Calculate the pH of the buffer system after 0.10 moles NaOH are added. Assume that there is no volume change during the additon of the NaOH.

25. The production and functioning of a buffer system has a limit. What is that limit?

COLLIGATIVE PROPERTIES

• Colligative Properties • Molality • Freezing Point •
• Boiling Point • Mole Fraction •
• Vapor Pressure • Osmotic Pressure •
• Questions and Problems •

COLLIGATIVE PROPERTIES

A colligative property is a characteristic of a pure solvent. The property of the solvent is modified by a solute dissolved in the solvent. Additionally, the change in the property of the solvent is increased by an increase in concentration of the solute.

The simplified version is that pure solvents, such as water, have specific physical characteristics (boiling point, freezing point, etc.). Each of these physical characteristics is modified when the water is not pure. The extent of the change in the characteristic is based on the amount of the solute that is dissolved in the solvent. An example is that pure water freezes at 0°C, but salt water does not freeze at 0°C. Also, the stronger the salt solution, the further away from 0°C is the freezing point of the salt solution.

The relationship between concentration and the change in a colligative property can be expressed by a mathematical formula. All we have to do is use the appropriate type of concentration in calculating the change in the colligative property. Concentrations are calculated in molarity, mole fraction, or osmolarity, depending on the colligative property.

MOLALITY

Molality (m) is a method of calculating concentration. As with molarity (M), molality starts with the moles of the solute. The difference between molarity and molality is that molarity is calculated by the number of moles per liter of solution; molality is calculated by the number of moles per kilogram of solvent. The methods of mixing are different, and the contents of each milliliter of solution are different.

FORMULA	$M = \dfrac{\text{moles solute}}{\text{L solution}}$	$m = \dfrac{\text{moles solute}}{\text{kg solvent}}$
TO MIX	1. Measure the number of moles required.	
	2. Dilute to number of liters required.	2. Dilute using number of kilograms solvent.
FINAL VOL.	Always known	Not known

Although we start at the same place in mixing the two solutions (moles solute), we do not know the effect of dissolving the solute in the solvent on the final volume. Furthermore, the weight of a solvent (kilograms solvent for molality) is tied to the density of the solvent for calculations that tell us the volume of the solvent used. All of this means that the molal solutions, unless they are weak, may not have a predictable final volume. On the other hand, molar solutions are mixed to a specific volume. The contents of every portion of that volume is known.

Example 12.1: What is the process for mixing a 2-m sucrose solution when you start with one mole of sucrose, $C_{12}H_{22}O_{11}$?

Solution: We know that the calculation requires a formula. A formula is nothing more than a set of instructions to lead to the answer. We write the formula, substitute y into the formula (y is the unknown value), and solve for the answer to this problem.

$$m = \frac{\text{moles solute}}{\text{kg solvent}}$$

$$2\ m = \frac{1\ \text{mole solute}}{y}$$

Let's substitute in the units of molality and cross multiply.

$$\frac{2\ \text{moles solute}}{1\ \text{kg solvent}} \diagdown\!\!\!\!\diagup \frac{1\ \text{mole solute}}{y}$$

Cross multiplication resolves this equation to a single line. The division through to isolate y shows us that we can cancel units.

$$2\ \text{moles solute} \times y = 1\ \text{kg solvent} \times 1\ \text{mole solute}$$

$$y = \frac{1\ \text{kg solvent} \times 1\ \cancel{\text{mole solute}}}{2\ \cancel{\text{moles solute}}}$$

Notice that the value of y is in the units of kg solvent. We need this value because we cannot complete the calculation without the number of kilograms of solvent.

$$y = 0.5\ \text{kg solvent}$$

We can produce the mixture by dissolving 1 mole sucrose in 0.5 kg of the solvent.

Example 12.1 does not mention the specific solvent name or formula. Under the terms of molality, kilograms of solvent is used, regardless of what the solvent really is. If the solution in Example 12.1 were mixed using water as a solvent, the volume of the solution would be different than if we used ethyl alcohol as the solvent. However, the solvent is assumed to be water when no solvent is named.

Example 12.2: A solution is mixed by dissolving 5.8 grams of table salt in one liter of water. What is the molality of this solution?

Solution: This problem can be solved by expanding the formula for molality with the substitution of what a mole is ($^g/_{MW}$) for moles solute (MW = molecular weight). We also use the density of water, 1 g/mL, to calculate the kilograms of solvent.

$$m = \frac{\text{moles solute}}{\text{kg solvent}}$$

$$m = \frac{\text{grams solute}/\text{MW solute}}{\text{kg solvent}}$$

We substitute in the value of MW, which is grams/1 mol.

$$m = \frac{5.8\ \text{g solute} \Big/ 58\ \text{g}/1\ \text{mol}}{1\ \text{kg solvent}} = 0.1\ \text{m NaCl}$$

The final volume of the solution produced in Example 12.2 is really not known. We suspect that the small amount of salt dissolved does not change the volume. If that is true, this solution can be handled as if it were a molar solution. However, we do not know the final volume until we mix the solution. Also, we normally assume that the density of water is $1.0 \, ^g/_{mL}$. Actually, the density of water is a little below 1.0 g/mL and does vary with temperature. At 25°C, room conditions, the density of water is $0.99707 \, ^g/_{mL}$, according to the *Handbook of Chemistry and Physics*. Therefore, 1 liter does not weigh exactly 1 kilogram.

FREEZING POINT

The effect of a solute on the freezing point (FP) of a solvent is to depress the freezing point. The amount of the freezing point drop is directly related to the molal concentration of *particles* in solution. If the solute is a nonelectrolyte, the number of particles is the same as the number of molecules. If the solute is an electrolyte, the number of particles per molecule depends on the number of ions released on disassociation and the extent of the disassociation.

Of course, there is a mathematical expression for all of this.

$$\Delta T_f = i K_f m$$

ΔT_f is the freezing point depression (the change in the freezing point).

i is the number of particles per molecule of solute.

K_f represents the constant for each of the solvent; each solvent has a specific K_f.

m is the concentration in terms of molality.

Once the calculation has been performed, ΔT_f is *subtracted* from the freezing point of the pure solvent to give us the freezing point depression.

$$FP_{solution} = FP_{solvent} - \Delta T_f$$

NOTE: The next section details boiling point changes. The calculations are handled in a similar manner to those for freezing point changes.

HINT: Many texts present problems that work with electrolytes as a separate topic. One way to avoid confusion is to consider whether or not the solute is an electrolyte in all problems that calculate a change in either freezing or boiling point.

Example 12.3: Calculate the FP of a solution composed of 150 g $C_6H_{12}O_6$ dissolved in 1 liter of water. Assume: $D = 1.0 \, ^g/_{mL}$; $MW = 180 \, ^g/_{mole}$.

Solution: The value of i for organic solutes is usually considered as 1. K_f for water is 1.86°C/m. The value of m is calculated from the information about the solution. We notice that we can express m by expanding to moles/kg.

$$\Delta T_f = i K_f m$$

$$\Delta T_f = 1 \times 1.86° \, C/m \times \frac{(150/180) \, mol}{1 \, kg}$$

$$\Delta T_f = 1.55°C$$

The freezing point of the solution is calculated.

$$FP_{solution} = FP_{solvent} - \Delta T_f$$

$$FP_{solution} = 0°C - 1.55°C$$

$$FP_{solution} = -1.55°C$$

Example 12.3 contains a solute that does not ionize. The next example addresses a solute that does ionize.

Example 12.4: What is the temperature at which the solution described in Example 12.2 freezes?

Solution: The solution from Example 12.2 is 0.1 m NaCl. Sodium compounds are very soluble in water and ionize completely. We need to take a look at the formula of the compound, NaCl, and apply the ions released to the value of i. NaCl disassociates to Na^{+1} and Cl^{-1} ions ($i = 2$).

$$\Delta T_f = iK_f m$$

$$\Delta T_f = 2 \times 1.86 \ °C/m \times 0.1 \ m$$

$$\Delta T_f = 0.372°C$$

We calculate the freezing point of the solution.

$$FP_{solution} = FP_{water} - \Delta T_f$$

$$FP_{solution} = 0°C - 0.372°C$$

$$FP_{solution} = -0.372°C$$

Not all solutions use water as a solvent. Benzene is an excellent solvent commonly used to dissolve organic solutes. Since the solutes are organic and organic solutes generally do not disassociate (or the disassociation is extremely weak), the value of i is one.

Example 12.5: What is the freezing point of a solution of 50 g of methanol (CH_3OH; MW = 32 $^g/_{mole}$) dissolved in 1.5 kg of benzene? The K_f for benzene is 5.12°C/m and the freezing point is 5.48°C.

Solution: We know that we need to substitute for m, moles. We can divide the mass of the methanol by the molecular weight of methanol to obtain 1.5625 of moles.

$$\Delta T_f = iK_f m$$

$$\Delta T_f = 1 \times 5.12°C/m \times \frac{1.5625 \ mol}{1.5 \ kg}$$

$$\Delta T_f = 5.33°C$$

The freezing point of the solution is calculated using the freezing point depression.

$$FP_{solution} = FP_{benzene} - \Delta T_f$$

$$FP_{solution} = 5.48°C - 5.33°C$$

$$FP_{solution} = 0.15°C$$

As was pointed out earlier in this chapter, we often do not know the final volume of a solution. Notice that the solution in Example 12.5 is produced by mixing benzene and methanol. Both of these substances are liquids. When we mix the two liquids, we end up with a solution that is certainly of a greater volume than the benzene alone. One of the factors that makes it particularly difficult to determine the final volume is that the density of each of these liquids is not 1 g/mL, and we do not know how the two substances interact in contributing to the final density of the mixture. We have no way to predict the density of the final solution, short of an empirical determination.

Colligative properties can be used to determine the molecular weight of a solute. The procedure is to weigh a sample of the unknown solute and dissolve in a known weight of the solvent. The freezing point of the solution is determined empirically. The information we have is sufficient to calculate the molecular weight of the sample when we use the expanded formula for ΔT_f. We expand the formula by substituting *g solute/g MW solute* for *m*.

$$\Delta T_f = iK_f m \quad \text{and} \quad m = \text{g solute/g MW solute}$$

After the substitution

$$\Delta T_f = \frac{i \times K_f \times \text{g solute/g MW solute}}{\text{kg solvent}}$$

We know all of the information for the ΔT_f calculation with the exception of gram MW solute, the requested answer.

Example 12.6: Fifteen grams of a substance determined to be a nonelectrolyte are dissolved in 100 grams of benzene. The freezing point of the solution is determined to be -2.0°C. Calculate the molecular weight of the unknown substance.

Solution: The substance is known to be a nonelectrolyte, which tells us that the value of *y* is 1 (no disassociation). We are given the mass of the benzene in grams; we need the mass in kilograms (100 g = 0.100 kg). The K_f and FP for benzene are found in Example 12.5 ($K_f = 5.12°C/m$ and FP = 5.48°C). The value of ΔT_f is the drop in FP from the normal FP (5.48 – –2.0 = 7.48°C change in FP).

We can either substitute directly into the ΔT_f equation or solve for gram MW solute and substitute. The equation has a complex fraction on the right. It is easier for most students to manipulate the symbols, rather than to manipulate the substituted numbers and units.

$$\Delta T_f = \frac{i \times K_f \times \text{g solute}/\text{g/MW solute}}{\text{kg solvent}}$$

Cross multiply.

$$\Delta T_f \times \text{kg solvent} = \frac{i \times K_f \times \text{g solute}}{\text{g/MW solute}}$$

Cross multiply again.

$$\Delta T_f \times \text{kg solvent} \times \text{g /MW solute} = i \times K_f \times \text{g solute}$$

Isolate g/MW solute.

$$\text{g/MW solute} = \frac{i \times K_f \times \text{g solute}}{\Delta T_f \times \text{kg solvent}}$$

Substitute in the values from the problem statement and calculate the answer.

$$\text{g/MW solute} = \frac{1 \times 5.12 \times 15}{7.48 \times 0.100}$$

$$\text{g/MW solute} = 102.7$$

Then the molecular weight of the unknown solute is 102.7 grams.

BOILING POINT

The effect of a solute on the boiling point (BP) of a solvent is to elevate the boiling point. The amount of the boiling point increase is directly related to the molal concentration of particles. Does this explanation for boiling point changes sound familiar? It is the same explanation that was used in the opening statements of the previous section dealing with freezing point depression. The reason for the duplication is that freezing point and boiling point changes follow basically the same pattern.

As with freezing point changes, we are still concerned about the values of i, m, and ΔT. However, the K_f and K_b are different; therefore, the values of ΔT_f and ΔT_b are different for the same molal concentration of solute. Additionally, the value of ΔT_b is added to the boiling point of the solvent, rather than subtracted, as in FP calculations.

$$\Delta T_b = iK_b m$$
$$\text{and}$$
$$BP_{solution} = BP_{solvent} + \Delta T_b$$

The value of K_b for water is 0.512°C/m. The K_b for water is presented because water is a very commonly used solvent.

Example 12.7: Calculate the boiling point for the glucose solution from Example 12.3. Compare the results in Example 12.4 and this example, Example 12.7.

Solution: The value of i for organic solutes is usually considered as 1. K_b for water is 0.512 °C/m. The value of m is calculated from the information about the solution (150 g solute in 1 kg solvent). We notice that we can express m by expanding to moles per kilogram (mol/kg) and cancel m against mol/kg.

$$\Delta T_b = iK_b m$$

$$\Delta T_b = 1 \times 0.512°\text{C/m} \times \frac{(150/180)\;\text{mol}}{1\;\text{kg}}$$

$$\Delta T_b = 0.43°\text{C}$$

The boiling point of the solution is calculated.

$$BP_{solution} = BP_{solvent} + \Delta T_b$$

$$BP_{solution} = 100°\text{C} + 0.43°\text{C}$$

$$BP_{solution} = 100.43°\text{C}$$

The comparison of this problem with Example 12.7 shows that the freezing point (0.372°C) change is greater than the boiling point change (0.43°C). The reason is the difference in the K_f and K_b; all other variables are the same in the two examples.

Example 12.8: Calculate the BP of the solution described in Example 12.5. The values required for benzene are K_b = 2.53°C/m and BP = 80.1°C.

Solution: From Example 12.5, we know that the molality is 1.5625 mol/1.5 Kg.

$$\Delta T_b = iK_b m$$

$$\Delta T_b = 1 \times 2.53°C/\text{m} \times \frac{1.5625 \text{ mol}}{1.5 \text{ kg}}$$

$$\Delta T_b = 2.6°C$$

The BP of the solution is calculated using the BP change, ΔT_b.

$$\text{BP}_{\text{solution}} = \text{BP}_{\text{benzene}} - \Delta T_b$$

$$\text{BP}_{\text{solution}} = 80.1°C + 2.6°C$$

$$\text{BP}_{\text{solution}} = 82.7°C$$

Example 12.9: A 25-g sample of a nonelectrolyte is dissolved in 175 grams of benzene. The boiling point of the solution is determined to be 83.0°C. Calculate the molecular weight of the substance.

Solution: This example is similar to Example 12.6. The differences are that the given information relates to BP, rather than FP; the substance weight is different; and the solute weight is different.

This substance is a nonelectrolyte; the value of i is 1. ΔT_b is determined by subtracting the BP of benzene from the BP of the solution ($\Delta T_b = 83.0 - 80.1 = 2.9°C$).

$$\Delta T_b = i \times k_b \times \frac{g \text{ solute}/g/\text{MW solute}}{\text{kg solvent}}$$

Cross multiply.

$$\Delta T_b \times \text{kg solvent} = i \times K_b \times \frac{g \text{ solute}}{g/\text{MW solute}}$$

Cross multiply again.

$$\Delta T_b \times \text{kg solvent} \times g/\text{MW solute} = i \times K_b \times g \text{ solute}.$$

Isolate g/MW solute.

$$g/\text{MW solute} = \frac{i \times K_b \times g \text{ solute}}{\Delta T_b \times \text{kg solvent}}$$

Substitute in the values from the problem statement and calculate the answer.

$$g/\text{MW solute} = \frac{1 \times 2.53 \times 25}{2.9 \times 0.175}$$

$$g/\text{MW solute} = 124.6$$

The molecular weight of the unknown solute is 124.6 grams.

MOLE FRACTION

Mole fraction is a method of expressing the concentration in moles of each component of a mixture relative to the total moles in the mixture. Generally speaking, the mole fraction is the target substance in moles as the numerator of a fraction (the top) divided by the sum of all the moles of the substances in the mixture. Mole fraction does not have units; it is dimensionless.

Suppose we have a mixture of two substances (A and B). We can express the mole fraction of substance A (X_A) in relation to the entire mixture.

$$X_A = \frac{\text{moles A}}{\text{moles A} + \text{moles B}}$$

The mole fraction of substance B (X_B) can also be expressed in the same manner.

$$X_B = \frac{\text{moles B}}{\text{moles A} + \text{moles B}}$$

Example 12.10: Calculate the mole fraction of NaCl in a mixture of 1.5 moles NaCl and 2.3 moles NH_4Cl.

Solution: We set up the formula for calculating the mole fraction, substitute in the numbers of moles, and solve for the answer mole fraction of NaCl.

$$X_{NaCl} = \frac{\text{moles NaCl}}{\text{moles NaCl} + \text{moles } NH_4Cl} = \frac{1.5}{1.5 + 2.3}$$

$$X_{NaCl} = 0.39$$

Example 12.11: What is the mole fraction of sucrose ($C_{12}H_{22}O_{11}$) in a mixture of 100 grams sucrose and 50 grams table salt (NaCl) dissolved in 0.5 L water? Assume whole number molecular weights.

Solution: We calculate the number of moles of each of the components of the mixture.

Sucrose: $12 \times 12 + 22 \times 1 + 11 \times 16 = 342 \ ^g/_{mole} \ C_{12}H_{22}O_{11}$
Table salt: $1 \times 23 + 1 \times 35 = 58 \ ^g/_{mole} \ NaCl$
Water: $2 \times 1 + 1 \times 16 = 18 \ ^g/_{mole} \ H_2O$

The next step is to calculate the number of moles of each component.

For sucrose
moles = $^{g \ sucrose}/_{MW}$
moles = $^{100 \ g}/_{342 \ g}$
0.29 moles sucrose

For table salt
moles = $^{g \ salt}/_{MW}$
moles = $^{50 \ g}/_{58 \ g/mole}$
0.86 moles salt

For water
moles = $^{g \ water}/_{MW}$
moles = $^{500 \ g}/_{18 \ g/mole}$
27.8 moles water

The final step is to set up the mole fraction formula, substitute in the numbers of moles, and solve for the desired answer ($X_{sucrose}$).

$$X_{sucrose} = \frac{\text{moles sucrose}}{\text{moles sucrose} + \text{moles salt} + \text{moles water}}$$

$$X_{sucrose} = \frac{0.29}{0.29 + 0.86 + 27.8}$$

$$X_{sucrose} = 0.010$$

Example 12.12: Calculate the mole fraction ($X_{solvent}$) in Example 12.11.

Solution: We can use the information in the solution to Example 11.11 and simply change from moles sucrose in the top of the fraction to moles water (the solvent).

$$X_{water} = \frac{\text{moles sucrose}}{\text{moles sucrose} + \text{moles salt} + \text{moles water}}$$

$$X_{sucrose} = \frac{27.8}{0.29 + 0.86 + 27.8}$$

$$X_{water} = 0.960$$

VAPOR PRESSURE

Experiments with nonvolatile solutions have found that the vapor pressure of the solvent is less than the vapor pressure of the pure solvent. Furthermore, the depression in vapor pressure is exclusive of the formula of the solute. The number of moles of the solute is the key to the vapor pressure depression. We can express the vapor pressure change by a simple relationship of vapor pressures (VP).

$$\Delta VP = VP_{pure\ solvent} - VP_{solution}$$

Raoult's law describes the vapor pressure depression in dilute solutions of nonvolatile nonelectrolytes as *the vapor pressure lowering of a solvent is proportional to the mole fraction of the solute present.* However, we can calculate the vapor pressure of the solvent directly from the mole fraction of the solvent.

$$P_{solvent} = X_{solvent} \times P^o_{solvent}$$

$P_{solvent}$ is the vapor pressure of the solvent as modified by the nature of the mixture. $X_{solvent}$ is the mole fraction of the solvent in the solution. $P^o_{solvent}$ is the vapor pressure of the pure solvent.

Vapor pressure lowering of the solvent is an explanation of why a solution boils at a higher temperature than the pure solvent. The boiling point of a solvent (or solution) is the temperature at which the vapor pressure equals the ambient (surrounding) vapor pressure. Since the vapor pressure of a solution is depressed, a higher temperature is required to equal ambient pressure; the boiling point is raised.

Example 12.13: The vapor pressure of water at 25°C is 23.76 torr. Calculate the vapor pressure at 25°C of a solution that contains 100 g sucrose in 1,000 g water.

Solution: This problem is a little different from Example 12.11. We are not told to use whole numbers. Instead, we use the complete weights from the Periodic Table of the Elements to calculate the molecular weights we substitute into the formulas for mole fraction and vapor pressure.

We calculate the molecular weights of sucrose and water.

For sucrose, $C_{12}H_{22}O_{11}$
$12 \times 12.011 + 22 \times 1.0079 + 11 \times 15.9994$
$MW_{sucrose} = 342.2992\ ^g/_{mol\ sucrose}$

For water, H_2O
$2 \times 1.0079 + 1 \times 15.9994$
$MW_{water} = 18.0152\ ^g/_{mol\ water}$

We calculate the moles of sucrose and water in the solution.

$100\ g/342.2992\ ^g/_{mol\ sucrose}$
0.292142 mol sucrose

$1000\ g/18.0152\ ^g/_{mol\ water}$
55.5087 mol water

We calculate the mole fraction of the water (X_{water}).

$$X_{water} = \frac{moles\ water}{moles\ sucrose\ +\ moles\ water}$$

$$X_{water} = \frac{55.5087}{0.292142 + 55.5087}$$

$$X_{water} = 0.9948$$

We substitute into the formula for the vapor pressure of the solution and calculate the vapor pressure of the solution.

$$P_{solvent} = X_{solvent} \times P^o_{solvent}$$

$$P_{solvent} = 0.9948 \times 23.76 \text{ torr}$$

$$P_{solvent} = 23.64 \text{ torr}$$

Although it was not requested by Example 12.13, we can calculate the change in the vapor pressure (ΔVP).

$$\Delta VP = VP_{water} - VP_{solution}$$

$$\Delta VP = 23.76 \text{ torr} - 23.64 \text{ torr}$$

$$\Delta VP = 0.12 \text{ torr}$$

OSMOTIC PRESSURE

OSMOSIS is defined as *the tendency of molecules to move from a greater concentration to a lesser concentration through a semipermeable membrane.* The membranes are capable of allowing specific substances through and not others. Osmosis is a very important mechanism useful in understanding the functions of living systems.

Since the movement of the molecules is predictable, from higher concentration to lower, there is a pressure factor involved (osmotic pressure, π). The osmotic pressure varies directly with the molarity of the solution. The osmotic pressure for dilute solutions of nonvolatile, nonelectrolytes is calculated using only two variables and one constant.

$$\pi = MRT$$

M is the molarity of the solution, R is the gas constant, and T is the temperature in the Kelvin scale. Because the gas constant is in atmospheres, the value of π is in atmospheres.

Example 12.14: Calculate the osmotic pressure of one liter of a solution that contains 100 grams of sucrose and is at 37.00°C (your body temperature).

Solution: We determine the molarity of the solution using the expanded version of the formula for molarity:

$$M = \frac{\text{moles solute}}{\text{L solution}} = \frac{\text{grams solute}/\text{grams per mol}}{\text{L solution}}$$

$$M = \frac{100 \text{ grams sucrose}/342.2992 \text{ grams per mole}}{1 \text{ L solution}}$$

$$M = 0.29214 \text{ }^{moles}/_L$$

The value of the gas constant, R, is $0.0821 \text{ }^{L \text{ atm}}/_{mole \text{ K}}$.

The temperature required for the osmotic pressure formula is in the Kelvin Scale because of the units associated with R. We are given 37.00°C and covert to the Kelvin scale to match the units in the gas constant.

$$K = °C + 273.15$$

$$K = 37.00 + 273.15$$

$$K = 310.15$$

We substitute into the formula for osmotic pressure and solve.

$$\pi = MRT$$

$$\pi = 0.29214 \;^{\text{moles}}/_{\text{L}} \times 0.0821 \;^{\text{L atm}}/_{\text{mole K}} \times 310.15 \text{K}$$

$$\pi = 7.44 \text{ atm}$$

QUESTIONS AND PROBLEMS

1. (a) What is the molality of a solution produced by dissolving 85 grams of glucose, $C_6H_{12}O_6$, in one liter of water?
    (b) Calculate the freezing point of this solution.

2. (a) Calculate the molal concentration of 2.5 liters of water in which 150 grams of a table sugar, $C_{12}H_{22}O_{11}$, is dissolved.
    (b) What is the freezing point of this solution?

3. (a) A 35.0-g sample of fructose, $C_6H_{12}O_6$, is dissolved in 275 g acetic acid. Calculate the molality of the solution. $(\Delta T_f = 3.90°C/m; FP = 16.6°C)$
    (b) Calculate the freezing point of this solution.

4. (a) What is the molality of a solution produced by dissolving 2.50 kg of fructose in 100 liters of water?
    (b) What is the boiling point of this solution?

5. (a) What is the concentration (molal) of a solution composed of 45 grams glucose and 750 grams ethyl alcohol, C_2H_5OH? $(\Delta T_b = 1.22\,C/m; BP = 78.5°C)$
    (b) What is the boiling point of this solution?

6. (a) Calculate the molality of a solution produced from 1.5 grams of NaCl dissolved in 77 mL of water.
    (b) What is the freezing point of this solution?
    (c) What is the boiling point of this solution?

7. A solution is produced by dissolving 45 grams of aluminum sulfate in 3 liters of water. Calculate the freezing point of the solution.

8. What is the boiling point of a solution composed of 18 grams of $C_3H_6O_3$ in 800 grams of nitrobenzene? Nitrobenzene boils at 210.88°C and has a molal boiling point elevation constant of 5.24.

9. Calculate the mole fraction of glucose, $C_6H_{12}O_6$, in a mixture of 50 g glucose dissolved in 250 mL H_2O.

10. What is the mole fraction of table sugar in a solution composed of 50 grams of sucrose $(C_{12}H_{22}O_{11})$ dissolved in 250 milliliters of water?

11. Notice that the given information in problems 9 and 10 is the same: 50 grams solute per 250 mL solute. Explain why the answers are different.

12. Calculate the vapor pressure at 30°C of 32 grams of glucose dissolved in 0.30 liters of water. (VP_{water} = 31.824 mm Hg at 30°C)
    REMINDER: 1 mm Hg = 1 torr of pressure.

13. The vapor pressure of water at 100°C is 760 mm Hg, the normal boiling point. One liter of water mixed with 50 grams of table sugar will not boil at 100°C. Why?

14. Calculate the osmotic pressure of a 3-M glucose solution at 25°C.

15. What is the osmotic pressure at normal human body temperature of a solution of 1.5 grams of a nonelectrolyte with a molecular weight of 135 diluted to 100 mL?

16. Calculate the molecular weight of the solute when the vapor pressure of a solution containing 7.5 grams of that nonvolatile nonelectrolyte dissolved in 85 grams of water is 22.98 mm Hg. These data were collected at 25°C. The vapor pressure for pure water at 25°C is 23.76 mm Hg.

17. A solution boils at 102.3°C. Analysis of the components of the solution tells us that there are 150 grams of a nonelectrolyte dissolved in 650 grams of water. Calculate the molecular weight of the solute.

18. A solution is produced by dissolving 56 grams of an unknown nonelectrolyte in 100 grams of water. The freezing point of the solution is –3.45°C. Calculate the molecular weight of the solute.

NUCLEAR CHEMISTRY

• Isotopes • Nuclear Decay • Half-Life •
• Determining Artifact Age • Effects of Exposure to Radiation •
• Measuring Emissions • Medical Applications •
• Questions and Problems •

ISOTOPES

Legend says that Henri Becquerel discovered the first of the radioactive substances by accident in 1895. He had a rock containing uranium on top a photographic plate and found that it was fogged when it was developed. Since 1895, there have been a number of naturally occurring radioactive elements discovered and artificially produced. The term **RADIOACTIVITY** was coined by Marie Curie.

Harold Urey discovered the existence of **HEAVY WATER**. Heavy water had a molecular weight higher than the predicted 18 grams per mole and was radioactive. This line of research led to the identification of **DEUTERIUM**, an isotope of hydrogen. This form of hydrogen had an atomic mass of 2, rather than 1. Urey's later studies resulted in the separation of radioactive isotopes from the commonly occurring elements of carbon, oxygen, nitrogen, and sulfur. Some of the other accomplisments are mentioned in Table 13.1.

1895	Wilhelm Roentgen	Discovered X-rays (1901, 1st Nobel Prize in physics)
1895	Jean Perrin	Performed cathode tube experiments leading to the discovery of the electron
1897	J. J. Thompson	Confirmed existence of, and measured the electron
1896	Henri Becquerel	Discovered radioactivity
1898	Marie Curie	Isolated polonium from uranium ore
1913	Niels Bohr	Proposed the structure of the hydrogen atom (Nobel Prize, 1922)
1911	Ernest Rutherford	Demonstrated positive charge in nucleus; proposed the neutron
1931	Harold Urey	Discovered heavy water (Nobel Prize, 1934)
1932	James Chadwick	Demonstrated existence of the neutron

Table 13.1
A brief listing of discoveries.

The Periodic Table of the Elements defines the specific element by means of the atomic number. The atomic number details the number of protons in the nucleus. The number of protons is equal to the number of electrons in a specific atom. However, the number of neutrons may be variable. Isotopes are different forms of an element due to differing numbers of neutrons present.

Deuterium, discovered by Urey, is one of three isotopes of hydrogen, as was discussed in Chapter 2. The three isotopes are hydrogen, deuterium, and tritium. Tritium is radioactive; hydrogen and deuterium are not radioactive. We refer to the radioactive isotopes by calling them **RADIOISOTOPES**.

It turns out that there are fewer isotopes of the lighter atoms than there are of the heavier. For instance, there are 8 isotopes of nitrogen, 9 of oxygen, 10 of phosphorous, 13 of chromium, and 22 of gallium. The lighter elements tend to have few isotopes whereas the heavier elements tend to have more isotopes. There are over 275 stable isotopes, and there are in excess of 800 radioisotopes.

When we consider the nonradioactive isotopes against radioactive isotopes, there is a relationship between the number of protons (Z, the atomic number) and the number of neutrons (N). In those isotopes with a ratio at or near $Z : N$, the isotopes tend to be stable. If the ratio shows that there are less protons than neutrons, $Z < N$, the isotopes tend to be unstable (radioisotopes). There are elements that have more than one stable isotope. $^{10}_{5}B$ and $^{11}_{5}B$ are both stable isotopes of boron. They both occur naturally in any sample of boron. The other four isotopes of boron are unstable; they are radioisotopes.

One method of referring to a specific isotope is to indicate the atomic mass after the symbol of the element. For example, we can use the boron isotopes and indicate by means of B-10 and B-11.

NUCLEAR DECAY

Radioisotopes are unstable. This means that the nucleus changes in numbers of either protons or neutrons or both. The process of nuclear change is called **DECAY**.

One of the particles that is emitted during decay is an **ALPHA PARTICLE**, α. An alpha particle is a helium nucleus $^{4}_{2}He$. Uranium's most common isotope, U-238, decays in a predictable manner. We write this **ALPHA DECAY** process in the same way that we write a chemical reaction.

$$^{238}_{92}U \rightarrow ^{234}_{90}Th + ^{4}_{2}He$$

Notice that the superscripts balance (238 amu against 234 amu + 4 amu). The superscripts are the atomic masses of the participants in the reaction, and mass must be conserved by the law of conservation of matter (we cannot gain or lose matter). In the case of the uranium decay reaction, the subscripts also balance. There are conditions under which the subscripts don't balance, as is discussed with beta decay later in this chapter.

The speed with which an alpha particle leaves the atom is rather high. Alpha particles travel at 10,000,000 m/s and faster. Since the particle is rather large, it is easy to top. Aluminum foil will stop alpha particles and so will a few sheets of notebook paper. Of course, the technician will not be dressed in aluminum foil or notebook paper; a heavy cotton lab coat will do the same job. The problem with exposure to alpha emission is that the DNA of the skin cells can be damaged, leading to either death of the cells or mutation of the genetic material. If mutation occurs, there is a chance of the exposure leading to skin cancer.

Example 13.1: Write the alpha decay reaction of Po-218, atomic number 84.

Solution: Nuclear reactions are written the same way as chemical reactions. We write the symbol, atomic number, and atomic weight (mass) for polonium on the left. We write a symbol, X, for the unknown atom produced by the decay. We record the complete symbol for a beta particle.

$$_{84}^{218}\text{Po} \rightarrow {}_{?}^{?}\text{X} + {}_{2}^{4}\text{He}$$

Recognizing that the law of conservation of mass applied, we subtract the mass of the alpha particle from the mass of the polonium and place the new mass by the new element, *X*. In the same manner we subtract the protons (2) in an alpha particle from the atomic number of polonium and record the new atomic number with *X*.

$$_{84}^{218}\text{Po} \rightarrow {}_{82}^{214}\text{X} + {}_{2}^{4}\text{He}$$

All that is left to do is identify the element with the atomic number of 82. The Periodic Table of the Elements tells us that *X* is lead, Pb.

$$_{84}^{218}\text{Po} \rightarrow {}_{82}^{214}\text{Pb} + {}_{2}^{4}\text{He}$$

The normal isotope (most common) for lead is Pb-207. The isotope produced by the alpha decay of polonium is somewhat heavier, Pb-218.

A second emission from decay is a **BETA PARTICLE**, β. Beta particles are much smaller than alpha particles. A beta particle is a high-speed electron, β^-, or a **POSITRON**, β^+, leaving the nucleus. The positron is, for all practical purposes, identical to an electron, except that the electron has a negative charge and the positron has a positive charge.

We write the **BETA DECAY** reaction in the same manner as the alpha decay. Let's take a look at the beta decay of an isotope of rubidium, Rb-87.

$$_{37}^{87}\text{Rb} \rightarrow {}_{38}^{87}\text{Sr} + \beta^-$$

This reaction is very different from the reaction shown for alpha decay. Notice that the atomic number goes from 37 (rubidium) on the left to 38 (strontium) on the right, but that the masses do not change. The reason is that a neutron can decay to a proton and a beta particle.

$$\text{n} \rightarrow \text{p} + \beta^-$$

This decay makes sense when we recall that a proton has a +1 charge and the beta particle, an electron, has a charge of -1. When they are closely associated, the overall charge of the union of the two particles is zero. We know that the charge on a neutron is also zero.

The interpretation of the reaction is that one of the neutrons in Rb decays to a proton. This increases the atomic number by one unit, and since the atomic number determines the atom, the new atom is the next atomic number, Sr. An electron has a very low mass in comparison to a proton or a neutron. Protons and neutrons weigh nearly the same mass, so there is no significant change in the mass.

Beta particles are much smaller than alpha particles, and they travel at a much higher speed. Because of their small size and the high speed, beta particles are harder to stop and will penetrate deeper into tissue. The alpha particle is stopped by a few layers of skin cells, but the beta particle can penetrate further into the skin layer. The damage from beta exposure is similar to that of alpha particles. Heavy clothing protects from both alpha and beta emissions. Of course, eye protection must be worn, and there can be no exposed skin.

Even though alpha and beta particles cannot damage organs when the exposure is from an external source, there are potential problems beyond skin damage. If an alpha or beta emitter is either inhaled or swallowed, the emissions can damage the lungs or the digestive system's tissue. Since there is the potential for DNA damage, mutation in particular, the risk of cancer is rather high.

NOTE: You might see similar symbols to those presented in this book. Notice that the symbol for the beta particle has a minus sign as a superscript—some sources omit the sign. Omitting the sign assumes that beta particles are only electrons and does not include the existence of positrons.

The list of symbols for the beta particle (electron) that are equivalent is as follows:

$$\beta = \beta^- = {}_{-1}^{0}e = {}_{-1}^{0}e^- = e^-$$

The list of symbols for the positron is much shorter:

$$\beta^+ = e^+$$

Example 13.2: Write the beta decay reaction of C-14.

Solution: The fact that we are dealing with beta decay tells us that we utilize the general reaction for beta decay.

$$n \rightarrow p + \beta^-$$

This means that we are losing one of the neutrons from the carbon, but the carbon nucleus is gaining a proton. Since we lose a neutron but gain a proton, there is no change in the atomic weight. However, the gain of the proton does give us an atomic number one unit higher.

We write the reaction in the same manner as we did in Example 13.1.

$$^{14}_{6}C \rightarrow {}^{14}_{7}X + \beta^-$$

All we do to identify the element X is refer to the Periodic Table.

$$^{14}_{6}C \rightarrow {}^{14}_{7}N + \beta^-$$

Thus, the result of beta decay of C-14 results in N-14, which is stable.

There is another form of beta decay that emits a positron rather than an electron. A positron is an **ANTIMATTER** particle. No, this is not science fiction. Antimatter particles are identical to subatomic particles (proton, electron, etc.), but they have opposite characteristics (charge sign, magnetic characteristics, etc.).

The mechanism for this reaction is the decay of a proton.

$$p \rightarrow n + \beta^+$$

This method of decay drops the atomic number by one unit because of the effective loss of one proton. However, the atomic weight remains the same because a neutron replaces the proton lost in the atomic mass calculation.

Example 13.3: What is the product of the positron emission from the nucleus of an astatine atom, As-208?

Solution: This example asks us to recall the reaction of proton decay with the production of a positron.

$$p \rightarrow n + \beta^+$$

We write the reaction for the decay of As-208 with no mass change. We do expect a one-unit drop in atomic number because of the proton that decays.

$$^{208}_{85}\text{As} \rightarrow {}^{208}_{84}\text{X} + \beta^+$$

A quick reference to the Periodic Table tells us that the element produced is an isotope of polonium, atomic number 84.

$$^{208}_{85}\text{As} \rightarrow {}^{208}_{84}\text{Po} + \beta^+$$

NOTE: There has been no mention of the effect of being radioactive on the chemical activity of an isotope. Radioisotopes react chemically in much the same manner as isotopes that are not radioactive. In other words, a radioisotope of carbon (C-14) reacts in the same way as a non-radioactive isotope of carbon (C-12).

Nuclear decay is often accompanied by gamma (γ) radiation. Sometimes the nuclear components become excited sufficiently during both alpha and beta decay that the excess energy is emitted as electromagnetic radiation. Gamma radiation is a form of electromagnetic radiation. Since gamma radiation is not a particle, there is no change in either the atomic number or the atomic weight. Gamma radiation should be indicated within the equation for the decay, but often it is not.

Gamma radiation is much more difficult to stop than are either alpha or beta particles. The particles can be stopped with heavy clothing. Gamma radiation can pass through the human body. Therefore, special high-density barriers are used, such as lead shielding or a few feet of reinforced concrete. Specialized suits can be worn for short periods of time, but are they not completely effective.

HALF-LIFE

Radioisotopes decay at a constant rate. However, nearly all radioisotopes decay at different rates. **HALF-LIFE** is the amount of time it takes for half of the nuclei in a sample to decay.

Oxygen-15 is a radioisotope. O-15 has a half-life of approximately 2 minutes (122 seconds). If we have 10 grams of O-15 right now, we predict that 5 grams exist after 2 minutes, 2.5 grams exist after 2 more minutes (4 minutes total), and 1.25 grams exist 6 minutes from now. It would not take even one hour until there is an extremely small amount of O-15 present. Table 13.2 presents the results of measuring the O-15 present through 10 half-lives.

Present mass O-15		10.00 g			
1st half-life	2 min	5.00 g	6th half-life	12 min	0.16 g
2nd half-life	4 min	2.50 g	7th half-life	14 min	0.08 g
3rd half-life	6 min	1.25 g	8th half-life	16 min	0.04 g
4th half-life	8 min	0.63 g	9th half-life	18 min	0.02 g
5th half-life	10 min	0.31 g	10th half-life	20 min	0.01 g

Table 13.2
Oxygen-15 decay.

The decay of all elements occurs in the same way as that of oxygen-15. The only difference is the amount of time of the half-life itself. Half-lives vary from a fraction of a second to beyond billions of years.

Yb-176	11.4 s
Tc-99m	6 hours
I-123	13 hours
Sr-93	7.5 months
Sr-90	29 years
Ra-236	1600 years
C-14	5700 years
U-238	4.5×10^9 yr.

Table 13.3
Half-lives of selected radioisotopes.

Example 13.4: Nitrogen-17 is a radioisotope, a beta emitter, with a half-life of approximately 4.2 seconds. If we start with 23 grams of N-17, how many half-lives does it take to decay sufficient N-17 to have less than 0.001 g N-17?

Solution: We can set up a table, like Table 13.2, which tells us the effect of each half-life on the sample.

Present mass N-17		23.00 g			
1st half-life	4.2 s	11.50 g	7th half-life	29.4 s	0.18 g
2nd half-life	8.4 s	5.75 g	8th half-life	33.6 s	0.09 g
3rd half-life	12.6 s	2.88 g	9th half-life	37.8 s	0.05 g
4th half-life	16.8 s	1.44 g	10th half-life	42.0 s	0.02 g
5th half-life	21.0 s	0.72 g	11th half-life	46.2 s	0.01 g
6th half-life	25.2 s	0.36 g	12th half-life	50.4 s	0.006 g

Our table tells us that the mass of N-17 drops below the 0.001 grams requested by the problem after 46.2 seconds. This means that the original 23 grams of N-17 decayed to less than 0.001 grams in less than one minute.

Example 13.5: Using the information from Example 13.4 and Table 13.3, how much time does it take strontium-90 to decay from 23 grams to less than 0.001 grams?

Solution: The number of half-lives we determined in Example 13.4 is an interval of time. For Sr-90, each of the half-lives is an interval of 29 years. All we have to do is multiply the time of each half-life, 29 years, by the number of half-lives.

11 half-lives × 29 years = 319 years

The interpretation of this answer is that it takes in excess of 319 years for 23 grams of Sr-90 to decay to less than 0.001 grams of Sr-90.

Notice that both N-17 and Sr-90 decay by the schedule of half-lives. The difference is the duration of the half-life.

There is a problem with Sr-90, however. Strontium is a product of nuclear fall-out from an atomic bomb and is radioactive (beta emitter).

Notice that strontium is directly below calcium on the Periodic Table of the Elements. The location on the Periodic Table indicates that strontium reacts chemically in much the same manner as calcium, but it is chemically more active than is calcium. This means that a chemical reaction that would normally use calcium will use strontium when strontium and calcium are both present.

Suppose cows are grazing in a field on which there is some fall-out. The milk produced would contain some strontium compounds instead of the normal calcium compounds. Anyone drinking that milk stands a very real chance of being exposed to emissions as long as that strontium is within the body.

Here are the last pieces to the puzzle. Bone contains calcium phosphate in rather large amounts. Sr-90 has a half-life of 29 years. There is a lot of potential for damage to DNA, possibly leading to cancer of bone, internal organs, and other systems.

DETERMINING ARTIFACT AGE

One application of half-life is in determining the age of a bone, seashell, pottery, or some other material found during investigations, such as those of archeology, paleontology, and anthropology. A commonly used technique is **CARBON DATING (RADIOCARBON DATING)**, which makes use of the decay of carbon-14. C-14 and the more common C-12 are found in a definite ratio naturally. The ratio is approximately 1 C-14 atom for each 1.3×10^{12} C-12 atoms present; this is a one to one trillion, 300 billion ratio. Carbon-14 decays at the rate of approximately 5,700 years per half-life.

Carbon-14 is produced in the upper atmosphere at a steady rate by the capture of a neutron by nitrogen. This process is driven by cosmic rays.

$$^{14}_{7}N + ^{1}_{0}n \rightarrow ^{14}_{6}C + ^{1}_{1}H$$

The C-14 reacts with atmospheric oxygen to produce carbon dioxide. The CO_2 enters the food chain via plant uptake. The nuclear decay of radioisotopes is not affected by the atom being in a chemical compound. The C-14 decays right on schedule as a beta emitter (and electron).

$$^{14}_{6}C \rightarrow ^{14}_{7}N + \beta^{-}$$

A problem with carbon dating is the small amount of C-14 found even in a very fresh sample. With the passage of time, the amounts of C-14 present become very difficult to detect. The measurements are exacting, requiring sophisticated equipment for accurate results. Even with the best equipment, this dating technique yields a reliable age of less than seven half-lives ($\pm 40,000$ years).

A way to extend the ability to date a sample is to choose an isotope with a longer half-life. An example is the use of K-40 dating.

$$^{40}_{19}K \rightarrow ^{40}_{19}Ar + \beta^{-}$$

This potassium radioisotope has a half-life in the range of one and a quarter billion years, with the decay product being argon. The potassium dating technique is ideal for determining the age of rock samples because of their age.

The Table of the Isotopes, found in the *Handbook of Chemistry and Physics*[1],

---

[1] *Handbook of Chemistry and Physics*, Editor-in-Chief Robert C. Weast, Ph.D., 66th edition (Boca Raton: CRC Press, Inc., 1985), B-233 ff.

presents the various isotopes and radioisotopes with the half-lives and other data. Our investigation of this table tells us that there are many naturally occurring radioisotopes that lend themselves to dating materials due to a sufficiently long half-life.

Example 13.6: A bone is found during an archeological expedition. A C-14 analysis is performed. The C-14 content of the bone is expected to be 0.00025 g, but it is found to be 0.00003125 g. What is the approximate age of the bone?

Solution: This bone's age can be approximated by determining how many half-lives have gone by. We divide the original mass by 2 and that mass by 2, etc.

Original mass C-14	0.00025 g	After first half-life	0.000125 g
		After second half-life	0.0000625
		After third half-life	0.00003125 g

This bone appears to be approximately three half-lives old. Each half-life represents approximately 5,700 years; therefore, the bone was formed in the range of 17,100 years ago.

Example 13.7: Another bone found in the same region is analyzed and found to contain 0.0000655 g C-14. What is the age of this second bone?

Solution: This bone's C-14 content is slightly over amount of C-14 expected after two half-lives and a bit less than the amount predicted for one half-life. Since the mass of C-14 in the sample is closer to the amount after the second half-life, we could express the estimated age of the bone as a time range. The bone can be considered to be just less in age than the 11,400 years of two half-lives. Another interpretation is that the age of the bone is between 8,550 and 11,400 years.

EFFECTS OF EXPOSURE TO RADIATION

DNA molecules are located in the nucleus of most cells. The function of DNA molecules is to carry information for the control of cell functions. This information is stored in the DNA molecule by means of the way in which the DNA molecule is put together.

Imagine that the alpha and beta particles are very small bullets traveling at high speeds. Also imagine what would happen if one of those bullets hit a DNA molecule, a long, complex molecule. The chances are that the DNA molecule would be broken, if not shattered. The pieces from the DNA molecule are polar and tend to come back together in a different manner than the original DNA molecule's structure. This change in the structure of DNA and the information carried by DNA is a **MUTATION**.

X-rays and gamma rays also cause mutations. These packages of energy tend to shatter the DNA molecule. A difference between the emissions is that the particles do not penetrate beyond the skin, but the rays can penetrate through the body if they have sufficient energy.

The first effect of a mutation may be absolutely nothing. Suppose the portion of the DNA molecule that is changed has no instructions for the functioning of the cell. If the new version of the DNA also has no instructions for the functioning of the cell, then there is no effect.

Another effect of the mutation is based on damage to information that controls cell function. If the mutation carries information that has no function, that function is lost. For instance, the absorption of vitamin B_{12} is through the walls of the gastrointestinal tract. The ability to absorb the vitamin is controlled by one's genetics. Pernicious anemia is the result of not being able to absorb vitamin B_{12}. The effects include extremely large red blood cells and, possibly, damage to the spinal cord.

The third effect is the change in information for a particular function to one that is still a function but is different. Suppose that the original information tells red blood cells that they are supposed to have the normal donut shape, but the mutation tells the cell to grow in a half-moon, elongated shape like a sickle. Cells of this shape tend to clump together, like a log jam, especially in the smaller blood vessels (capillaries). Among the symptoms are loss of organ function, pain in the joints, and low energy levels. This is a form of anemia, sickle-cell anemia.

The effects of mutation just discussed are why it is so important to protect those tissues that produce gametes, the sperm and ova. If any mutation occurs to gametes, the children produced will contain that mutation.

A person exposed to sufficient a dose of X-ray or gamma emissions can experience the signs of **RADIATION SICKNESS** due to gastrointestinal and/or bone marrow damage. The various symptoms of radiation sickness, also called **ACUTE RADIATION SYNDROME**, are loss of appetite, nausea, vomiting, abdominal pain, fever, diarrhea, hair loss, susceptibility to infection, and hemorrhage. There is a high incidence of dehydration. Radiation sickness is avoided by shielding, appropriate clothing, radiation detectors, and respect for emitters in office/hospital/industrial settings.

Ultraviolet (UV) light is particularly damaging to skin cells and the cells of the retina at the rear of the eye. A mutation that occurs in skin cells produces in excess of a half million cases of nonmelanoma skin cancer yearly. Ultraviolet light is highly suspected as the cause of melanoma, which can metastasize if not caught early. Since UV light is stopped with clothing and/or a sun blocker, UV is the easiest radiation type from which skin can be protected. Eye protection is in the form of sunglasses worn outdoors.

X-rays are commonly used in the diagnosis of medical and dental problems. X-rays penetrate tissue very well and can damage that tissue, as can any mutagenic emitter. The normal view is that the benefits of the exposure far outweigh the risks. Even so, care is taken to make certain that the gamete-producing tissues and the sensitive mammary tissues are not exposed (a lead-cloth apron is worn during a dental X-ray) and that the exposure is of the least intensity that will do the job. Of course, the individual performing the X-ray makes certain that he or she does not receive a dose—stepping behind a shield during the exposure and wearing a film badge to measure accumulated exposure.

MEASURING EMISSIONS

Exposure to emissions carries a potential danger that cannot be ignored. People who work in an environment that includes X-rays and alpha, beta, or gamma emitters can protect themselves to a degree by monitoring their own exposure.

The most common method of measuring personal exposure is a film badge (Fig. 13.1) pinned on the outside of clothing by those who work in areas where X-ray machines and emitters are located. The film badge is worn during the entire workday and is submitted periodically to have the film from inside the plastic carrier developed. Thefrequency of the film check is determined by the chances of being exposed on the basis of the nature of the equipment/emitter, the extent of shielding, and the volume of the equipment/emitter usage. Unexposed film appears to be transparent with no fogging. Exposed film is cloudy; the degree of exposure is detected by the density of the cloudiness.

Figure 15.1
Film badge.

There are emission detectors, some of which are electronic in nature and too large to wear. A Geiger counter or a scintillation counter, both of which are portable, give an instant reading of exposure by means of reading a meter and/or listening to a sound generator (clicking or a whistling).

Of course, there are various units for expressing the quantity of the emission being measured. The first two of the following descriptions are the physically related units; the remainder are used to describe possible effects on living tissues.

One **CURIE** (Ci) is the number of disintegrations of one gram of radium per second. This is 37,000,000,000 disintegrations per second.

A **BECQUEREL** (Bq) is defined as one disintegration per second. This unit is not often used in the medical area but is more appropriate in the field of nuclear physics.

The **ROENTGEN** (R) is directly related to the effects of the exposure of living tissue. One roentgen is the production of 2,100,000,000 ion pairs in 1 cm^3 of dry air at room temperature and pressure. R's are used to measure X-rays and/or gamma rays.

One roentgen and one rad are very close to the same value (1 R = 0.96 rad) and, therefore, they can be considered as being equivalent for practical purposes. One **RAD** (D) is the transfer of 2,400 calories of energy to 1 kg of tissue. One **GRAY** (Gy) is 100 times larger than a rad—the transfer of 240,000 calories of energy to 1 kg of tissue.

The rad and the rem are considered to be the standards when discussing the degree of danger from radiation exposure. The **RAD** (**R**adiation **A**bsorbed **D**ose) is equivalent to 0.01 gray (2.4 calories of energy transferred to 1 g of tissue). This appears to be a very small amount of exposure; however, the effect can be large when you consider the possible damage to the DNA in that tissue.

An exposure of 50 rads or less brings about no apparent health problem, but gamete-producing tissue may be affected. Most people will become ill with radiation sickness if they receive 200 rads, but they will survive. Four hundred fifty rads is sufficient to cause the death of approximately half of the humans exposed. Generally speaking, no one can survive an exposure of 800 rads.

The **REM** (**R**adiation **E**quivalent, **M**an) takes into account the effect of exposure of specific tissues found in humans to one rad of X-rays. The other units are generic; the rem is not. Natural radiation from radioisotopes found in the soil and other portions of the environment, plus radiation from the Sun, accounts for less than 0.5 rem per year. The federal guidelines for medicine and industry are 5 rems per year or less. The value of the rem is calculated by multiplying rad by a quality factor. The quality factor varies upward with the ability to produce ions; therefore, it is different for each source of radiation and is also determined by the purity of the source.

Example 13.8: A film badge is developed that was worn by an X-ray technician. The analysis of the developed film indicates that the technician received an exposure of 0.1 rad during the period of one week. What is the effect of this exposure on the technician?

Solution: An exposure of 0.1 rad over one week is not serious when considered as an isolated incident. However, the long-term exposure picture is a totally different story. If we assume a 50-week work year and that a $0.1 \, ^{rad}/_{week}$ exposure is normal for this individual, this technician might receive a total exposure of 5 rad annually.

$$0.1 \, \text{rad}/_{\cancel{week}} \times 50 \, \cancel{\text{weeks}}/\text{year} = 5 \, ^{rad}/_{year}$$

An exposure of $5 \, ^{rad}/_{year}$ is well below the level where there is concern, 50 rads.

Example 13.9: A reading is taken in a room in which researchers work with radioisotopes. It is determined that an exposure of 0.12 roentgens over the period of 3 hours is the risk to the workers. The policy of the research facility is that no person shall receive more than 2 rads as a maximum dose before a mandatory two-week rest period away from the job and a complete physical examination. How long can one work in this environment?

Solution: This company acknowledges that exposure is cumulative. We know that 1 roentgen is 0.96 rad; we have a conversion factor of $^{1 \, R}/_{0.96 \, rad}$ (or $^{0.96 \, rad}/_{R}$) that can be applied to change the roentgens to rads.

The easiest way to work this problem is to determine the number of roentgens per hour risk. Then we convert this to rads per hour. We know that the maximum allowed is 2 rads and we can determine how many hours it takes to receive a 2-rad dose.

$$\frac{0.12 \, \text{R}}{3 \, \text{h}} = \frac{0.04 \, \text{R}}{1 \, \text{h}}$$

We notice that 0.04 R/h can be expressed as 1 h/0.04 R. If we use 1 h/4 rad, we can easily calculate the amount of time that it takes to receive one rad exposure.

$$\frac{1 \, \cancel{R}}{0.96 \, \text{rad}} \times \frac{1 \, \text{h}}{0.04 \, \cancel{R}} = \frac{26 \, \text{h}}{1 \, \text{rad}}$$

We now know that it takes 26 hours to receive an exposure of 1 rad. All we have to do is figure out how many hours it takes to receive a 2-rad exposure.

$$2 \, \cancel{\text{rad}} \times \frac{26 \, \text{h}}{\cancel{\text{rad}}} = 52 \, \text{hours}$$

NOTE: Exposure to radiation is a serious risk. Exposure is accounted for over the period of one year. This means that the effects of exposure are considered to be cumulative; the greater the exposure over time, the greater the chances are of damage.

MEDICAL APPLICATIONS

X-rays are used in the diagnosis of bone injury, dental status, intestinal tract problems, and other investigations. X-rays are effective in viewing of bony tissue. X-rays generally penetrate soft tissues, but these tissues can be examined with the use of a radio-opaque substance. (An example is the barium sulfate used to enhance intestinal X-rays.)

X-rays can also be used to treat skin cancers. The advantage of X-rays for this use is that the X-rays not only can be focused, but they do not penetrate as far as gamma rays. Then X-rays are very easy to control.

Another use for X-rays, especially effective in the treatment of cancer, is **HYPERBARIC**, the exposure of a patient to X-rays when the patient is at 3 atmospheres of pressure. Normal cells are much more susceptible to death under these circumstances than are healthy cells. This amounts to a selective method for killing cancer cells.

The thyroid gland is a major user of iodine for the production of thyroid hormone. If the thyroid gland is not functioning properly, the problem can be diagnosed by looking at the iodine uptake. The diagnostic method is to have the patient drink a sodium iodide solution. The sodium iodide is produced from a radioisotope of iodine, ^{131}I or ^{123}I. The thyroid gland is inspected by a scan of the neck or can even be performed by exposing black and white film with the emission from the iodine. The choice of isotope is related to the half-life and the emission. I-123 is a gamma emitter with a half-life of slightly over 13 hours, whereas I-131 is a beta and gamma emitter with a half-life of a bit over eight days. From the standpoint of long-term effects, I-123 has a shorter half-life and might make more sense than I-131.

Radioisotopes of iodine are also used to treat thyroid cancer since the iodine is concentrated in the thyroid gland. This localization of the iodine tends to reduce the exposure of the remainder of the body. I-131 might be the better choice due to the longer half-life.

Radioisotope	Half-Life	Medical Use
Cobalt-60 $^{60}_{27}$Co β γ	5.3 years	Treatment of cancer
Technetium-99 $^{99}_{43}$Tc γ	6 hours	Scans of brain, heart, liver, kidney, and full-body bone for structure and cancer
Strontium-85 $^{85}_{38}$Sr γ	64 days	Scan of bone
Phosphorous-32 $^{32}_{15}$P β	14.3 days	Treatment of leukemia
Iron-59 $^{59}_{26}$Fe β	45.1 days	Scan of bone marrow and determination of extent of anemia

Table 13.4
Selected radioisotopes and their uses.

Table 13.4 presents some of the radioisotopes commonly used in medicine. The choice of the radioisotope for medical diagnosis and/or treatment is based on half-life and emission types. An example of the choice process is detailed previously in the discussion of iodine isotopes. The best choices are those radioisotopes with a half-life that is short but is long enough to produce the test material and to perform the test or have the desired effect during treatment. Toxicity of the radioisotope and its product(s) is a major consideration because illness (or death) might accompany a toxic substance. Of course, the nature of the emission must be such that detection is easily possible; gamma rays are easier to detect than particles that may be stopped by tissues.

QUESTIONS AND PROBLEMS

1. Write the equation for the alpha decay of actinium-210, $^{210}_{89}\text{Ac}$.

2. Radon is a radioactive gas that can make its way into a building from the soil. What are the products of the alpha decay of $^{212}_{86}\text{Rn}$?

3. $^{116}_{47}\text{Ag}$ decays to $^{112}_{45}\text{Rh}$. What is the means of the decay?

4. Write the equation for the beta decay of thorium-231, $^{231}_{90}\text{Th}$.

5. Detail the beta decay of mercury-206.

6. What is the equation for the β^- decay of $^{214}_{82}\text{Pb}$?

7. What is the equation for the β^+ decay of $^{202}_{82}\text{Pb}$?

8. Iodine-131 can decay by means of positron decay. Write the equation that describes this decay process.

9. Write the equation for the decay of the positron emmiter, $^{131}_{54}\text{Xe}$.

10. What is the element (isotope) that is produced by means of an alpha decay of Xe-131, then a beta decay?

11. What is the isotope produced by a beta, beta decay of Po-209?

12. Starting with californium-251, what is the product of an alpha, alpha, beta decay?

13. An alpha, beta, beta, alpha decay is synthetically produced starting with the most common isotope of iron. What is the theoretical product?

14. What was the isotope that underwent a single beta decay to become Cd-110?

15. A beta decay is theorized to produce Cl-37. What is the isotope we start with?

16. Uranium-235 is produced from an element stimulated to undergo decay in which positron is emitted. What is the starting element?

17. The half-life of Pb-202 is 53,000 years. What is the length of time involved in 8 half-lives?

18. Silver-104 has a half-life of 69 months. How many years does it take for 0.15 grams of silver to decay to 0.0046875 grams?

19. What is the approximate number of half-lives that pass when 3 micrograms of a radioisotope decay to 0.003 micrograms?

20. There are two radioisotopes of iodine used in nuclear medicine. I-123 is a gamma emitter with a half-life of slightly over 13 hours. I-131 is a beta and gamma emitter whose half-life is approximately 8.1 days. If you had the choice, which would you use for treatment of cancer of the thyroid gland?

21. Tc-99 is used for diagnostic purposes for the brain, kidney, liver, bone, and cardiac muscle. The radioisotope is injected and the full body is scanned. Explain the appearance of the scan in a case of a localized cancer of the liver and the reason for performing a full-body scan.

22. Kr-81 is an inert gas and a gamma emitter. The half-life of Kr-81 is over twenty thousand years. Why is a radioisotope with such a long half-life used to diagnose lung ventilation?

23. An individual is exposed to 3 R over a period of 2 days. How long will it take for this person to receive a 50-R exposure?

24. How does the exposure described in problem 23 affect the individual when it is continued for one year?

25. In one week, an office worker is exposed to 2.3 roentgens due to a flaw in the shielding of the room in which diagnostic X-rays are performed. Calculate the exposure for one year, assuming that the shielding defect is not repaired. What are the risks to the office worker of this exposure?

HYDROCARBON FAMILIES

• Alkane Family • Alkene Family • Alkyne Family •
• Questions and Problems •

ALKANE FAMILY

Generally speaking, organic molecules are those that are carbon based. Figuring out how the participants in an organic molecule interact is similar to using the oxidation numbers, but we do not really need to worry about the charge on the oxidation number. The oxidation number tells us the number of chemical bonds that can form when a compound is produced. For instance, oxygen is a negative 2 in a compound; this means that oxygen forms two chemical bonds. Another example is that iron can have an oxidation number of positive 3, which indicates that iron can take part in three chemical bonds.

If we look at just the number of bonds that form, putting together organic molecules is just like putting together a jigsaw puzzle. We know what the pieces look like and place them so that there are no pointed ends (unused bonds) sticking out. Table 14.1 shows how the most common puzzle pieces appear.

Element	Number of Bonds
Carbon	4
Hydrogen	1
Oxygen	2
Sulfur	2
Nitrogen	3
Phosphorous	3

Table 14.1
Organic puzzle pieces.

The simplest of the organic molecules is methane, CH_4. Methane is a **HYDROCARBON**, a molecule composed of hydrogen and carbon. Methane is commonly known as marsh gas; it is an end product of the metabolism of some species of bacteria that typically live in the wet soils of marshes and swamps. The hydrogen-to-carbon-to-hydrogen bond angle in methane is 109°30' (one hundred nine degrees and thirty minutes, which is 109.5°). This bond angle is different if a hydrogen is replaced (substituted) with an ion or if there is a double bond between carbons.

Methane is a member of the **ALKANE** family of hydrocarbons. The alkanes are hydrocarbons chemically bound by one carbon-to-carbon bond. Since methane, however, only has one carbon, let's look at the next member of the alkane family, a two-carbon compound. Notice that the sketch of methane indicates that the angles are difficult to draw. Therefore, for convenience, sketches of organic molecules are usually drawn using a square pattern. This method of sketching is faster and gives us an easy way to visualize a two-dimensional version of the molecular structure.

Suppose we were to remove one of the hydrogens from two methane molecules.

We end up with two ions, or what organic chemists refer to as methyl groups. They can join to produce a two-carbon compound, ethane. The bond between the carbons is a single bond; therefore, ethane is an alkane.

$$\underset{\overset{\displaystyle |}{H}}{\overset{\overset{\displaystyle H}{|}}{H-C-}} \quad + \quad \underset{\overset{\displaystyle |}{H}}{\overset{\overset{\displaystyle H}{|}}{-C-H}} \longrightarrow \underset{\overset{\displaystyle |}{H}\;\overset{\displaystyle |}{H}}{\overset{\overset{\displaystyle H}{|}\;\overset{\displaystyle H}{|}}{H-C-C-H}}$$

. There is absolutely no reason why we cannot perform the same reaction to produce a three-carbon chain by substituting a methyl group, CH_3, for one of the hydrogens of an ethane. This sounds complicated, but it isn't. The hydrocarbon produced with the three-carbon chain is propane.

$$\underset{\overset{\displaystyle |}{H}}{\overset{\overset{\displaystyle H}{|}}{H-C-}} \quad + \quad \underset{\overset{\displaystyle |}{H}\;\overset{\displaystyle |}{H}}{\overset{\overset{\displaystyle H}{|}\;\overset{\displaystyle H}{|}}{-C-C-H}} \longrightarrow \underset{\overset{\displaystyle |}{H}\;\overset{\displaystyle |}{H}\;\overset{\displaystyle |}{H}}{\overset{\overset{\displaystyle H}{|}\;\overset{\displaystyle H}{|}\;\overset{\displaystyle H}{|}}{H-C-C-C-H}}$$

We can produce the remainder of the alkane family by the same technique of adding one methyl group each time. We build the next largest alkane each time we add a functional group. The four-carbon version is butane, and the five-carbon alkane is pentane. Starting with the five-carbon alkanes, we name by numerical designations. Table 14.2 outlines the names up to ten carbons.

1 C's	Methane	6 C's	Hexane
2 C's	Ethane	7 C's	Heptane
3 C's	Propane	8 C's	Octane
4 C's	Butane	9 C's	Nonane
5 C's	Pentane	10 C's	Decane

Table 14.2
Number of carbons and naming the alkanes.

Decane is far from the largest member of this family. We can keep adding carbons as methyl groups to existing chains and end up with some very large molecules.

Adding methyl groups is not the only way to perform the addition of carbons. If we remove one hydrogen from an ethane molecule, we end up with an ethyl group.

$$\underset{\overset{\displaystyle |}{H}\;\overset{\displaystyle |}{H}}{\overset{\overset{\displaystyle H}{|}\;\overset{\displaystyle H}{|}}{H-C-C-}}$$

We can remove one hydrogen from any of the alkanes and produce a group that we can place the same way as we did with the methyl group. We can produce large molecules very quickly using larger groups than the methyl group for substitution. The naming of each group is done by removing *ane* and placing *yl* as the suffix. Therefore, prop*ane* becomes the prop*yl* group.

Hydrocarbon group replacements of a hydrogen on an alkane are not the only substitutions possible. Nearly any ion, such as a fluoride ion, can substitute for a hydrogen. The product of the substitution is methyl fluoride (fluoromethane). Incidentally, the fluoride ion can be substituted for any of the hydrogens in methane since there is no difference between them.

$$H-\overset{\overset{\displaystyle H}{|}}{\underset{\underset{\displaystyle H}{|}}{C}}-F$$

Suppose we substitute more than one ion for the hydrogens on methane. For instance, suppose we substituted a bromide, a chloride, a fluoride, and an iodide and had to name the compound. The method of naming is to alphabetize the substituents and name the compound that was substituted. Notice that we cannot differentiate between the carbons bound to methane due to the tetrahedral shape of the molecule.

$$Br-\overset{\overset{\displaystyle F}{|}}{\underset{\underset{\displaystyle Cl}{|}}{C}}-I$$

Bromochlorofluoroiodomethane

On the other hand, we can substitute ethane and longer hydrocarbons in a number of ways. The result of the various substitutions does not change the general formula for the compound, but it does change the structural formula. All the different structural formulas that have the same general formula are **ISOMERS**. It is important to be able to tell the difference between specific isomers because they react chemically in different ways. We name isomers to indicate the nature of the structure.

The **TERMINAL CARBON** is the end carbon that has been most substituted or the end carbon on the side of the molecule that has the most substitutions. We can look at this problem in two different ways if we substitute two bromines for two hydrogens on the ethane molecule. We can replace a hydrogen with the bromines on the same carbon or on different carbons.

$$H-\overset{\overset{\displaystyle H}{|}}{\underset{\underset{\displaystyle H}{|}}{C}}-\overset{\overset{\displaystyle Br}{|}}{\underset{\underset{\displaystyle H}{|}}{C}}-Br \qquad\qquad H-\overset{\overset{\displaystyle Br}{|}}{\underset{\underset{\displaystyle H}{|}}{C}}-\overset{\overset{\displaystyle Br}{|}}{\underset{\underset{\displaystyle H}{|}}{C}}-H$$

1,1-dibromoethane   **1,2-dibromoethane**

Both of the preceding structural formulas, CH_3CHBr_2 and CH_2BrCH_2Br, have the same general formula, $C_2H_4Br_2$. The result of the different substitutions is called a pair of **ISOMERS**. Isomers are compounds with the same general formula but different structural formulas. We differentiate between the isomers by naming them to indicate the nature of the substitution.

We consider one of the terminal carbons (end carbon) to be the number 1 carbon. The terminal carbon that has had the most chemical activity or is nearest to the most active site is the number 1 carbon. The carbon next to the number 1 (terminal) carbon is the number 2 carbon, etc.

We name the number of bromines substituted and where they appear by using *the number of the carbons involved*. Additionally, we name *how many times the substituent is used* (which is the reason for the use of *dibromo* in the names). This naming procedure uses redundancy, but it helps us by clearly pointing out the total number of times each substituent is used.

Sometimes we think that we have drawn an isomer, but we must look at the sketch and use some imagination to make certain we are correct. For instance, let's start with propane and attempt to sketch the various isomers possible with a methyl group as the substituent. We can place a methyl group on a terminal carbon, or we can place a methyl group on the interior carbon. Note that placing a methyl group on the left terminal carbon is the same as placing one on the right terminal carbon.

Compound A   **Compound B**

Compound A and Compound B are not isomers. Compound A is really pentane because there are five carbons that form the straight carbon chain. This is easy to see if we draw compound A as a straight-chain structural formula. The straight-chain method of drawing molecules used is for convenience. Recall that the bond angles are not 90°, but they are really 109°30' when we have only carbon and hydrogen in the alkanes and there is one chemical bond between carbons. This means that the right hand terminal carbon in Compound A can be rotated, and the carbon skeleton for Compound A *is* really five carbons long. Compound A is pentane.

Compound B is a substituted propane. We name the most active terminal carbon as the number one carbon. In the case of this compound, the central carbon is the one on which the substitution occurs—there is no specific number 1 carbon. We can name either terminal carbon as the number 1 carbon. The central carbon, then, is the number 2 carbon. Compound B is 2-methylpropane.

Example 14.1: What is the IUPAC (International Union for Pure and Applied Chemistry) name for this compound?

Solution: The carbon skeleton is butane. In alphabetical order, the substituents are bromo, chloro, fluoro, and iodo. The left terminal carbon is the number 1 carbon. There is more substitution to the left of the geographic center of the compound.

The name of this compound is 1-bromo 2-chloro 2-fluoro 3-iodo butane.

Example 14.2: Sketch the substituted hydrocarbon, 2,2-dichloro 1-fluoro 3-iodo propane.

Solution: This compound is based on the three-carbon hydrocarbon, propane. Let's assign the left terminal carbon as the number 1 carbon. Once this assignment is made, all we have to do is place the substituents on the appropriate carbon. Notice that the compound turns out to be symetrical. This means that it makes no difference which terminal carbon we designate as the number 1 carbon.

$$\begin{array}{ccccc} & H & Cl & H & \\ & | & | & | & \\ F- & C- & C- & C- & I \\ & | & | & | & \\ & H & Cl & H & \end{array}$$

Example 14.3: How many isomers are there when a methyl group is substituted for one hydrogen belonging to pentane?

Solution: We already know that we cannot substitute the methyl group for a hydrogen on either of the terminal carbons. If we do that substitution, we end up with hexane. We can, however, substitute on any of the interior (circled) carbons.

$$\begin{array}{ccccccc} & H & H & H & H & H & \\ & | & | & | & | & | & \\ H- & C- & \textcircled{C} & \textcircled{C} & \textcircled{C} & C- & H \\ & | & | & | & | & | & \\ & H & H & H & H & H & \end{array}$$

There is one small problem with using the carbons on either side of the center-most carbon. Those carbons are in the same position. Imagine using the second carbon from the left as the number two carbon. Isn't that the same as the second carbon from the right if we flip the molecule 180° horizontally? Of course it is. Thus, the two isomers produced are based on substituting the methyl group on the central carbon and on either one of the carbons just to the left of the center. Let's substitute starting from left to right so we can have the left terminal carbon serve as the number 1 carbon.

2-methylpentane	3-methylpentane

Notice that in the case of 3-methylpentane, either of the terminal carbons can be the number 1 carbon. They are equally distant from the active site on the molecule, the central carbon.

ALKENE FAMILY

The **ALKENE** family compounds have a minimum of one double bond in the carbon skeleton of the members of the alkenes. Two hydrogens, one each from adjacent carbons, are removed.

The open chemical bonds from the two carbons allow the second bond to form between carbons. The nature of the bond is due to the sharing of electrons, a covalent bond.

The ethane molecule is normally sketched more than one way depending on the way we need to see it. We can draw ethane either in a three-dimensional representation or in one of the more convenient two-dimensional views (Fig. 14.1). The most common choice for the two-dimensional view is the square sketch because of the convenience and ease in drawing this view.

| Three Dimensional | Angular | Square |

Figure 14.1
Sketches of ethene.

As with the alkane family, various ions can be substituted for one (or more) of the hydrogens on the ethene molecule. If we place the two iodines on the same carbon, it does not matter which, the compound produced might be named *diiodoethene*. Even though the name looks descriptive, it can tell us more if we take a closer look at its components.

The first problem with the name *diiodoethene* is that we can substitute both iodines on one of the carbons, or we can place them on separate carbons. If we name the location of each iodine substitution, using the number of the carbon, we know the difference between substituting both iodines on one carbon or one iodine on each of the carbons. The two possibilities are *1,1-diiodoethene* if the two iodines are on the same carbon and *1,2-diiodoethene* if the two iodines are on one carbon each.

A second problem with the naming of compounds with multiple bonds on the carbon skeleton appears when there is more than one pair of carbons between which the bond can appear. We first recognize that the double bond is between the first and second carbon and, second, know that we are dealing with the alk*ene* family. We name the location of the double bond by using *1-ene*, which tells us that the bond begins on the number one carbon. We name the carbon skeleton, the two carbons, as if we were naming a member of the alkane family. The alkane name is used because the alkenes are derived from the alkanes by means of the removal of two hydrogens to produce the second bond between the carbons. Application of this information provides us with a very descriptive name, *1,1-diiodo 1-ene ethane*.

Naming the position of the double bond for the ethene molecule is not important. The problem occurs when we have more than two carbons in the skeleton of the molecules. The end of the molecule closest to the double bond becomes the number 1 carbon, and the remainder are numbered sequentially to the other terminal carbon. The carbons between the terminal carbons are the **INTERIOR CARBONS**.

Example 14.4: What is the name of the compound in this sketch?

$$\begin{array}{ccccc}
H & & & Br & \\
\backslash & & & | & \\
C & = & C & - & C & - H \\
/ & & | & & | \\
H & & H & & H \\
\end{array}$$

Solution: The order for writing the name is from right to left. We start with the name of the skeleton, any bonds that are double, and the substituents in alphabetical order.

We count the number of carbons in the carbon skeleton of the molecule (the longest straight chain). There are three carbons, which tells us that the compound is a modified propane molecule.

The number 1 carbon is determined by the position of the double bond. We name the location of the double bond by the number of the carbon from which the bond starts (the left terminal carbon is number the number 1 carbon), *1-ene* propane.

We locate the bromine by the number of the carbon on which it is located, the number 3 carbon, *3-bromo*.

We have the complete name: *3-bromo 1-ene propane.*

Example 14.5: What is the name of this compound?

$$\begin{array}{ccccccc}
H & & & Br & & H & \\
\backslash & & & | & & | & \\
C & = & C & - & C & - & C & - Cl \\
/ & & | & & | & & | \\
H & & H & & H & & H \\
\end{array}$$

Solution: The carbon skeleton is four carbons long. This means that we name this compound on the basis of butane. The double bond is on the left; the left-hand carbon is the number 1 carbon. The identification of the double-bond location is *1-ene* because the bond starts from the first carbon and extends to the second.

The substituents are bromine and chlorine in alphabetical order. The bromine is located on the number 3 carbon, *3-bromo*. The chlorine is substituted for a hydrogen on the number 4 carbon, *4-chloro*.

We put the name together to get *3-bromo 4-chloro 1-ene butane.*

There is absolutely no reason why we cannot draw the structural formula for a molecule if we are given the name. Take a look at Example 14.6.

Example 14.6: Sketch the structural formula for 3,4-dibromo 1-chloro 5-iodo 2-ene pentane.

Solution: The first step is to draw the carbon skeleton for pentane.

C–C–C–C–C

Let's use the right-hand carbon as the number 1 carbon. We place the double bond (2-ene) starting at the number 2 carbon and extending to the number 3 carbon.

C–C–C=C–C

The next substituent is iodine (5-iodo) in any location on the number 5 carbon.

I–C–C–C=C–C

We have one chlorine to place (1-chloro) on the number one carbon.

I–C–C–C=C–C–Cl

We place the two bromines (3,4-dibomo) on the number 3 and 4 carbons.

$$\begin{array}{c} Br \\ | \\ I\text{–}C\text{–}C\text{–}C\text{=}C\text{–}C\text{–}Cl \\ | \\ Br \end{array}$$

There are numerous chemical bonds that are not satisfied. Each of the carbons must have four chemical bonds. The remainder of the bonds not used above are occupied with hydrogens, and then we are finished.

$$\begin{array}{ccccccc} H & H & Br & H & H \\ | & | & | & | & | \\ I\text{--}C\text{--}C\text{--}C\text{=}C\text{--}C\text{--}Cl \\ | & | & & & | \\ H & Br & & & H \end{array}$$

Another factor to consider is that the double bond between the carbons is a rigid bond. If we use a ball-and-stick model to show the structure of ethene, we cannot use the sticks to join the carbons by a double bond. The bond angle should be 109°30', but this bond angle for the second bond is bent. The other factor involved is that single bonds allow for rotation, just like an airplane propeller. A double bond between carbons does not allow for this turning, called torsion. Since there is no torsion between carbons when there is a double bond, the hydrogens are not equivalent.

Suppose we substitute one iodine on each of the carbons in ethene. There are two ways of performing the substitution (two isomers) and, of course, we have a system for telling the difference between the two isomers by the names.

cis-1,2-diiodo 1-ene ethane   *trans*-1,2-diiodo 1-ene ethane

The only difference in the names is the use of *cis* and *trans*. *Cis* is used to indicate that the two iodines are on the same side of the double bond between the carbons. *Trans* tells us that the two iodines are located across the double bond.

Example 14.7: Name the substituted hydrocarbon.

$$\begin{array}{c} Cl \qquad\qquad Br \\ \backslash \qquad\qquad\ | \\ C\text{=}C\text{--}C\text{--}H \\ / \qquad\ |\quad | \\ H \qquad\ Cl \ \ H \end{array}$$

Solution: This compound is based on a propane molecule that has a double bond between the number 1 and 2 carbons, 1-ene propane. Alphabetically, we also have bromine and chlorine. The bromine is on the number 3 carbon, 3-bromo. The chlorines are one each on the number 1 and 2 carbons and are across the double bond, *trans*-1,2-dichloro.

We put the name together: 3-bromo *trans*-1,2-dichloro 2-ene propane

The **ALKYNE** family members have at least one triple bond between the carbons forming the skeleton of the molecule. The $H-C\equiv C-H$ simplest of the alkynes is acetylene, the gas that is used for welding.

The name *acetylene* is a **CLASSICAL NAME** (a name that is in general use). The IUPAC name should be *ethyne* because the two-carbon naming is by *eth*, as in ethane, and the word ending should be *yne* to indicate the triple bond between carbons. The name *acetylene* has been in use so long that it is not likely to be changed, regardless of IUPAC rules. Many classical names are used and are pointed out throughout the remainder of this book.

Example 14.8: Determine the IUPAC name of this substituted hydrocarbon.

$$
\begin{array}{ccccc}
 & F & & F & H \\
 & | & & | & | \\
H- & C- & C\equiv C- & C- & C-H \\
 & | & & | & | \\
 & H & & H & F
\end{array}
$$

Solution: This molecule has a five-carbon skeleton and is based on pentane. We see that the triple bond is to the left of the center of the molecule; the left-hand carbon is the number 1 carbon. The triple bond is named as **2-yne**. There are three fluorines to be named, 1,4,5-trifluoro.

The name of this compound is 1,4,5-trifluoro 2-yne pentane.

SHORTCUT: There are times when drawing the structural formula is not desired or practical. In these cases, we can write the formula for the compound and indicate the structure. The formula for the compound in Example 14.8 is CH₂FCCCHFCH₂F.

NOTE: Word endings (suffixes) are of particular importance in naming organic molecules. The suffix used indicates a great deal about the nature of the compound, as we will see in later discussions.

QUESTIONS AND PROBLEMS

1—4   Name these compounds.

1.
$$
\begin{array}{ccc}
Cl & Cl & Cl \\
| & | & | \\
H-C- & C- & C-H \\
| & | & | \\
H & I & H
\end{array}
$$

2.
$$
\begin{array}{ccccc}
H & I & H & H & F \\
| & | & | & | & | \\
H-C- & C- & C- & C- & C-F \\
| & | & | & | & | \\
H & H & Cl & H & H
\end{array}
$$

3.
$$
\begin{array}{cc}
H & H \\
| & | \\
H-C- & C-H \\
| & | \\
H & C \\
 & H\triangle H \\
 & H
\end{array}
$$

4.
$$
\begin{array}{cccc}
 & & H & \\
 & & H\,C\,H & \\
H & H & | & H \\
| & | & | & | \\
Cl-C- & C- & C- & C-Cl \\
| & | & | & | \\
H & H & H & H
\end{array}
$$

5—12   Draw these compounds.

5.   1-bromo 2,4-difluoro pentane

6.   1,1-dichloro 2-fluoro ethane

7.   1,2-diiodo 2,4,4-trifluoro pentane

8.   1,1,6,6-tetrachloro 2,2-difluoro 3,5-dimethyl hexane

9.   3,3,6-trichloro 4-ethyl octane

10.   1,1,2,4-tetrachloro 2,6,7-triiodo nonane

11.   2,6-dibromo 3-iodo 4,5-dimethyl 1-ene hexane

12.   6-bromo 5-chloro 1,3-diene octane

13—18   How many isomers are there of these compounds?

13.   Hexane substituted by one fluorine

14.   Ethyl hexane

15.   Methyl octane

16.   Ethane substituted with one fluorine and one chlorine

17.   Difluoro propene

18   *Trans*-diflouro hexene

19—20   Name these compounds.

19.

20.

ALCOHOLS, ACIDS, ESTERS, AND ETHERS

• Alcohols • Acids • Esters • Ethers •
• Questions and Problems •

ALCOHOLS

The **ALCOHOL** functional group in organic chemistry has the same structure as the hydroxide ion in general chemistry. The alcohol group is a puzzle piece with one open chemical bond, which is from the oxygen. The alcohol group is bound to a carbon, either terminal or interior. This carbon is sometimes called the alcohol or alcoholic carbon. Although the nature of the chemical bonds from the oxygen is angular, for convenience the bonds are often written as **–O–H**.

The simplest of the alcohols is obtained by substituting one of the hydrogens on a methane molecule with the alcohol functional group. The compound is called methanol or, using one of the older names, methyl alcohol. The common name for methanol is wood alcohol, which is a central nervous system toxin. Methanol poisoning can be fatal. One of the interesting uses for methanol is to get rid of water in a gas tank. Another use is as a solvent.

The two-carbon version of an alcohol is ethanol, also called ethyl alcohol. Ethanol's common name is grain alcohol; it can be produced by the action of yeasts on the carbohydrates found in grain. Does this sound familiar? Ethanol is found in all alcoholic drinks. Ethanol is a neurotoxin that affects the higher abilities and, in the extreme, can shut down the autonomic nervous system. Ethanol is also found in extracts (cinnamon, vanilla, maple, and others) as an organic solvent. Maple extract contains approximately 14% ethanol. Highly concentrated ethanol is used as a dehydrating agent.

An alcohol group can be substituted onto a propane molecule, but there is a bit of a problem. There is more than one way to perform the substitution. We can place the alcohol group on a terminal carbon OR we can place the alcohol group on an interior carbon, ending up with a pair of isomers. As with any substituted hydrocarbon, the carbon skeleton, propane, is then placed with the alcohol group, *ol*, by indicating the carbon on which it is substituted.

1-ol propane   **2-ol propane**

The classical names are most often used for these isomers. 1-ol propane is named normal propyl alcohol or *n*-propanol; 2-ol propane is called isopropyl alcohol (rubbing alcohol), or iso-propanol.

Example 15.1: How many isomers can be drawn using a pentane skeleton and one alcohol group?

Solution: For convenience, we can draw the carbon skeleton of pentane without the hydrogens.

$$\begin{array}{ccccc} H & H & H & H & H \\ | & | & | & | & | \\ H-C-C-C-C-C-H \\ | & | & | & | & | \\ H & H & H & H & H \end{array} = \begin{array}{ccccc} | & | & | & | & | \\ -C-C-C-C-C- \\ | & | & | & | & | \end{array}$$

We draw the molecule with the alcohol group placed in all possible locations, but we must be careful that there is no duplication.

1-ol pentane	2-ol pentane	3-ol pentane

Placing an alcohol group in any other location on pentane gives us a duplicate molecule. There are three isomers of pentanol.

A compound in which the alcohol group is bound to a terminal carbon is called a **PRIMARY ALCOHOL**; 1-ol pentane is a primary alcohol. If the alcohol is bound to an interior carbon, the bonding is referred to as a **SECONDARY ALCOHOL**, as are 2-ol pentane and 3-ol pentane.

ACIDS

Just as there is a functional group that behaves like a base, the alcohol group, there is a group that behaves like an acid. The **CARBOXYL CARBON** (boxed) fits the definition of an acid because a hydrogen ion can leave the group. One characteristic of organic acids is that they do not ionize well; they tend to be weak acids.

The sketch of the carboxyl carbon in the previous paragraph indicates the general appearance of the O-H component as angular. However, it is often inconvenient to draw this portion of the molecule in the angular manner. Therefore, as a general agreement among chemists, it is OK to straighten out the carbon-oxygen-hydrogen angle.

If we count the chemical bonds to the carboxyl carbon, we find there are four. If we try to place the carboxyl carbon as an interior carbon, we have five bonds. Carbons can only have four bonds; the carboxyl carbon is a terminal carbon.

> *NOTE: The sketch showing the carboxyl carbon includes a symbol,*
> ***R***. *This symbol is used to represent the remainder of the molecule,*
> *which can be just a hydrogen or a very large structure.*

The simplest of the acids is obtained if R is a hydrogen. This acid's name is *methanoic acid*. Notice that the compound is derived from meth*ane*. We name the acid by replacing the *e* in *ane* with *oic* as the word ending. (Methan*e* becomes methan*oic* acid.)

Methanoic acid is most often referred to by its classical name, formic acid. Formic acid is a product of some animals and is used as a toxin. Red ants (fire ants) produce formic acid and inject it when biting. The sensation is a stinging or burning, and some people are violently allergic (hypersensitive) to this toxin.

$$\underset{\displaystyle H-C-O-H}{\overset{\displaystyle O}{\overset{\displaystyle \|}{}}}$$

The two-carbon acid is produced if we substitute a methyl group for the *R*. This acid is one that most people are familiar with by its classical name, acetic acid. Another name (IUPAC) is ethanoic acid. A weak solution of approximately 5% acetic acid is vinegar. Stronger solutions of acetic acid are used in photographic film processing, including X-ray film, and in the developing of photographic prints.

It is possible to have an acid that contains more than one carboxyl carbon since organic molecules have more than one end. Malic acid is an acid with more than one carboxyl group. Citric acid is another example having more than one carboxyl carbon. As a matter of fact, there are three ends, each a carboxyl carbon.

Malic Acid   Citric Acid

ESTERS

If we think in terms of acid-base reactions, the products are a salt and water, as discussed in Chapter 11. Organic acids, bases, and alcohols behave in the same way. The acid loses the hydrogen H– from the carboxyl carbon, and the alcohol loses the alcohol group –OH with the final product of water. The remainder of the organic acid and alcohol bond to form the organic salt. The only difference is that an organic salt is called an **ESTER**.

The naming of an ester follows the same general rules as with an inorganic salt. The function group that comes from the alcohol, the base, is named first. The second portion of the name is the acid's name with the suffix *ate*.

Example 15.2: Write and balance the chemical reaction illustrating the reaction of formic acid and ethanol.

Solution: We write the structural formulas of formic acid and ethanol with the carboxyl carbon and alcohol group facing each other because they are the active groups in an acid-base reaction. We know from our discussions of inorganic and organic acid-base reactions that water is formed and indicate where the water comes from by circling the H and OH that are involved. The ester is produced from the remaining ions.

The name of the ester is derived from the alcohol's hydrocarbon group (ethyl), and then the acid's name is expressed as an *ate* (formate).

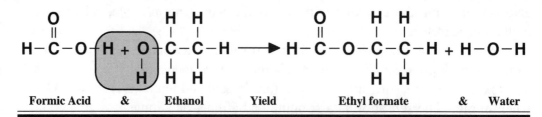

Formic Acid & Ethanol Yield Ethyl formate & Water

There are biologically important compounds that are generated by the reactions of acids and alcohols. For instance, **LIPIDS** can be isolated from both plant and animal tissue by the use of a nonpolar solvent such as benzene. The nature of lipids leads to the classification: Those that are solids at room temperature are **FATS** and those that are liquids at room temperature are **OILS**. A few examples of lipids are triglycerides (fats and oils), phospholipids from membranes, glycolipids.

Triglycerides, also called glycerol trialkanoates, triacylglycerols, and glycerides, are used by animals as an energy storage molecule. The triglycerides are esters of glycerol (glycerine) and a series of acids that are called **FATTY ACIDS**. These acids are composed of one carboxyl carbon, and most have long straight-chain hydrocarbon in place of R from the general formula for an acid. That straight-chain addition to the carboxyl carbon is most often composed of an uneven number of carbons.

The four-carbon fatty acid derived from butane is butanoic acid (classical name butyric acid) and is an example of one of the smaller of the fatty acids. It is composed of the usual single carboxyl carbon with a three-carbon straight chain (propyl group) replacing one of the carboxyl carbon's hydrogens.

Glycerol is a triol (three alcohol groups) and can react as a base in a chemical reaction with an acid. If we present one of the terminal carbons of glycerol with an acid, there is a reaction in which an ester linkage is formed.

Glycerol & Fatty Acid Yield Lipid & Water

There is no reason why this same reaction cannot occur on the remaining carbons of the glycerol molecule. At the completion of this reaction, we have a lipid and, of course, three water molecules. Since we have a lipid produced from three butyric acids, this lipid is often referred to as a tributyrate. This particular lipid is one of the lipids that are found in butterfat.

Is it necessary to have all three fatty acids the

same? No, it is not. There are quite a few naturally occurring and synthetic lipids that are produced from more than one fatty acid. Commercial vegetable oils can contain the esters of oleic acid, linoleic acid, and palmitoleic acid with glycerol in nearly any combination.

Notice that the carbon chain in the butyric acid molecule is held together by single bonds between carbons. This bonding of carbons allows for the maximum number of hydrogens on the molecules. The lipids produced from this fatty acid are **SATURATED** fats. **UNSATURATED** lipids are produced from fatty acids that contain one or more double bonds between carbons in the chain.

Oleic, linoleic, and palmitoleic acids are unsaturated; therefore, the lipids produced are liquids and classified as oils. Sometimes the unsaturated lipids are **HYDROGENATED**. This means that the hydrogens are placed back into the molecule and react at the most active points, which are the double bonds between carbons. The result of hydrogenation is to convert or partially convert an oil into a fat. An example of this process is the production of shortening from vegetable oils.

ETHERS

We can look at ethers in pretty much the same way as we handled the formation of esters. We produce esters from the reaction of an acid and a base, with one of the products being water. The production of ethers is much the same in the sense that water is produced along with the ether. The difference is that we start with two alcohols in the production of ethers. The general formula for an ether, R–O–R, is very similar to that of an ester except that the doubly bound oxygen found in carboxyl carbons is not present on the alcoholic carbon.

The reactions that produce esters and ethers are **DEHYDRATION** reactions and normally require a dehydrating agent. Concentrated sulfuric acid, H_2SO_4, is a good dehydrating agent and is commonly used in the reactions that yield ethers. The use of sulfuric acid (or whatever) is indicated in the reaction. Let's look at the production of an ether from the reaction of two methanols.

Methanol   Methanol   Yield   Methyl Ether   &   Water

The naming system for ethers is straightforward: Ethers are normally named by the common name. We name functional groups derived from alcohols by increasing complexity and indicate the nature of the linkage (bonding) with *ether*. The name for the ether we just produced can either be methyl ether or dimethyl ether.

Example 15.3: Name this ether.

Solution: This is a simple ether. These are usually named by the functional groups in order of complexity followed by *ether*. The ethyl group is more simple (two carbons) than is the propyl group (three carbons). The name of this ether is ethyl propyl ether.

Example 15.4: Produce the ether from the dehydration of methanol and ethanol.

Solution: We draw the alcohols, indicate the dehydration, and name the ether.

$$H-\overset{\overset{\displaystyle H}{|}}{\underset{\underset{\displaystyle H}{|}}{C}}-O \; + \; O-\overset{\overset{\displaystyle H}{|}}{\underset{\underset{\displaystyle H}{|}}{C}}-\overset{\overset{\displaystyle H}{|}}{\underset{\underset{\displaystyle H}{|}}{C}}-H \; \xrightarrow[\text{H}_2\text{SO}_4]{\text{Conc.}} \; H-\overset{\overset{\displaystyle H}{|}}{\underset{\underset{\displaystyle H}{|}}{C}}-O-\overset{\overset{\displaystyle H}{|}}{\underset{\underset{\displaystyle H}{|}}{C}}-\overset{\overset{\displaystyle H}{|}}{\underset{\underset{\displaystyle H}{|}}{C}}-H \; + \; H-O-H$$

Methanol   &   Ethanol   Yield   Methyl Ethyl Ether   &   Water

Diethyl ether (ethyl ethyl ether) is the earliest of the inhaled anesthetics. The anesthetic properties of diethyl ether were discovered accidentally by Dr. C. W. Long in 1842. This ether is fairly easy to control, but it is slow to act. Moreover, recovering patients commonly experience nausea and/or vomiting along with headaches. The major use of diethyl ether today is as an organic solvent used for extractions, especially for fats, oils, and other organic molecules.

$$H-\overset{\overset{\displaystyle H}{|}}{\underset{\underset{\displaystyle H}{|}}{C}}-\overset{\overset{\displaystyle H}{|}}{\underset{\underset{\displaystyle H}{|}}{C}}-O-\overset{\overset{\displaystyle H}{|}}{\underset{\underset{\displaystyle H}{|}}{C}}-\overset{\overset{\displaystyle H}{|}}{\underset{\underset{\displaystyle H}{|}}{C}}-H$$

A great deal of care must be used when we work with ethers. Most ethers, especially the small ones, are extremely volatile and flammable. Ethers can present a major fire or explosion hazard if not handled very carefully. Always use ethers in a well-ventilated area or in a fume hood, and replace the cap as soon as possible. Ethers, either the liquid or the fumes, cannot be exposed to pure oxygen, spark, or flame.

QUESTIONS AND PROBLEMS

1. The alcohol functional group is similar to which ion in fields of inorganic chemistry?

2. Alcohols do not ionize to any great extent. As an example, methanol ionizes very poorly, but sodium hydroxide ionizes very well. Propose an explanation.

3. What is the structural difference between a primary and a secondary alcohol?

4. Sketch 2-ol butane. What is the classification of this compound?

5. Provide the structural formula of 3-ol pentane and classify the compound.

6. Describe the structural difference between an alcohol and an acid.
   (*HINT: Produce a sketch of the general formulas.*)

7. Explain why the carboxyl carbon must be a terminal carbon.

8. Describe the chemical activity of an alcohol with an organic acid.

9. What is the chemical equation for the dehydration reaction of propanioc acid and 1-ol propane (*n*-propanol)?

10. Refer to problem 9: What is the class of compound produced? Name the compound produced.

11. 1-ol butane (*n*-butanol) and acetic acid react to produce a compound. Write the chemical reaction indicated.

12. What is the name of the compound produced in problem 11? What is the class of compounds produced?

13. What is the class of compounds to which this structure belongs? What is the chemical reaction that produced this compound? What is the name of the compound?

```
     H  O        H  H
     |  ||        |  |
 H — C — C — O — C — C — H
     |           |  |
     H           H  H
```

14. What is the name of this compound?

```
     H  H  H  H        O  H  H
     |  |  |  |        ||  |  |
 H — C — C — C — C — O — C — C — C — H
     |  |  |  |           |  |
     H  H  H  H           H  H
```

This compound is a member of which class of compounds? What are the reactants that produced this compound?

15. What is the type of chemical bonding (linkage) that produces lipids? What is the analogous reaction in inorganic chemistry?

16. Nitroglycerine is a medication used to stimulate the heart and increase coronary circulation. Nitroglycerine is produced from glycerine and nitric acid. Write the equation for this reaction using structural formulas.

17. What is the difference between a saturated fat and an unsaturated fat?

18. The list of ingredients on some food items includes "partially hydrogenated" oil. What does this mean?

19. Both alcohols and organic acids are polar. Sketch the general formula for each and indicate the location(s) of the negative polar points.

20. What is the structural difference between an ester linkage and an ether linkage? (*HINT: Sketch the general formulas, and then compare.*)

21. What is the name of this compound? Write the reaction for the production of this compound using structural formulas.

$$
\begin{array}{ccccccccc}
& H & H & & H & H & H \\
& | & | & & | & | & | \\
H- & C- & C- & O- & C- & C- & C- & H \\
& | & | & & | & | & | \\
& H & H & & H & H & H
\end{array}
$$

22. Name this compound and identify the compounds from which it is produced.

$$
\begin{array}{ccc}
H \\
| \\
H-C-H & H \\
| & | \\
H-C-O-C-H \\
| & | \\
H-C-H & H \\
| \\
H
\end{array}
$$

OTHER CLASSES OF ORGANIC COMPOUNDS

• Aldehydes • Ketones • Amines •
• Thiols • Aromatics • Heterocyclics •
• Questions and Problems •

ALDEHYDES

Imagine an aldehyde functional group as a carboxyl carbon that has the-OH portion removed and replaced with a hydrogen. The oxygen-carbon portion of the functional group is the **CARBONYL CARBON**.

$$R-\overset{\overset{\displaystyle O}{\|}}{C}-H$$

Carbons can only have four chemical bonds, and three are already taken by the doubly bound oxygen and the hydrogen. The fact that the *R* is placed on a carbon that already has three chemical bonds means that the aldehyde group is a terminal carbon.

An aldehyde is produced by the oxidation of a primary alcohol. An example is the oxidation of methanol that results in the formation of methaldehyde.

$$H-\overset{\overset{\displaystyle H}{|}}{\underset{\underset{\displaystyle H}{|}}{C}}-\overset{\overset{}{}}{\underset{\underset{\displaystyle H}{|}}{O}} + [O] \longrightarrow H-\overset{\overset{\displaystyle O}{\|}}{C}-H + H-O-H$$

Methanol & Oxygen Yield Formaldehyde & Water

We notice that the name *formaldehyde* is not an IUPAC name. This is the classical name and is not likely to be changed. The IUPAC name recognizes that the aldehyde is derived from methane. Furthermore, the suffix *al* is used to indicate that we are dealing with an aldehyde. The IUPAC name for this aldehyde is methanal.

Formaldehyde (methanal) is, of course, used as a preservative for animal tissues. Modern preservatives for animals used in dissections either contain little formaldehyde or none at all because formaldehyde infiltrates through the skin and does major bone damage. Even so, it is a good idea to wear gloves during dissections and to wash your hands thoroughly when you are finished.

Another aldehyde is acetaldehyde, CH_3CHO. This aldehyde is a neurotoxin and is the product of alcohol (ethanol) oxidation. Acetaldehyde (IUPAC name ethaldehyde) is oxidized to acetic acid, but the process is very slow. Once acetic acid has been formed, the metabolism to carbon dioxide and water is fast.

$$H-\overset{\overset{\displaystyle H}{|}}{\underset{\underset{\displaystyle H}{|}}{C}}-\overset{\overset{\displaystyle H}{|}}{\underset{\underset{\displaystyle H}{|}}{C}}-O + [O] \longrightarrow H-\overset{\overset{\displaystyle H}{|}}{\underset{\underset{\displaystyle H}{|}}{C}}-\overset{\overset{\displaystyle O}{\|}}{C}-H$$

$$H-\overset{\overset{\displaystyle H}{|}}{\underset{\underset{\displaystyle H}{|}}{C}}-\overset{\overset{\displaystyle O}{\|}}{C}-H + [O] \longrightarrow H-\overset{\overset{\displaystyle H}{|}}{\underset{\underset{\displaystyle H}{|}}{C}}-\overset{\overset{\displaystyle O}{\|}}{C}-O-H$$

Not all aldehydes cause problems; for instance, the flavor of cinnamon is an aldehyde, as is vanilla.

Example 16.1: Draw the structure and name the five-carbon aldehyde.

Solution: We sketch the five-carbon straight chain for convenience and indicate the carbonyl carbon as a terminal carbon.

$$
\begin{array}{ccccc}
H & H & H & H & O \\
| & | & | & | & \| \\
H-C-&C-&C-&C-&C-H \\
| & | & | & | & \\
H & H & H & H &
\end{array}
$$

The name of this compound is pentanal. An alternate name is pentyl aldehyde.

Example 16.2: An aldehyde exists that has a carbon skeleton that is eight carbons long. Name that compound.

Solution: An eight-carbon, straight chain is octane. The aldehyde is named octanal and can be called octyl aldehyde.

KETONES

Ketones are produced from the carbonyl group being placed in the interior of a compound rather than as a terminal carbon. This means that there is a doubly bound oxygen in the heart of the compound. The R groups indicated in the general formula to the right are hydrocarbon derivatives if we look at the compound in this manner.

$$
\begin{array}{c}
O \\
\| \\
R-C-R
\end{array}
$$

Acetone (propanone) is the first of the ketones because it is the smallest possible ketone. Acetone is based on the propane molecule with the carbonyl group in the only place it can be—it is the middle carbon. Acetone is a possible metabolic product in the metabolism of glucose, but acetone is

$$
\begin{array}{ccc}
H & O & H \\
| & \| & | \\
H-C-&C-&C-H \\
| & | & | \\
H & H & H
\end{array}
$$

produced in toxic amounts by individuals affected by diabetes mellitus. Acetone can be detected on the breath. Acetone is a neurotoxin that causes the person to have poor balance and slurred speech, as if this person were drunk. Detectable acetone is a symptom that the person needs professional aid immediately.

We can develop common names for ketones by identifying the R groups and naming them as a ketone. If we have an ethyl group and a propyl group tied together by the carbonyl group, we have a ketone.

$$
\begin{array}{cccccc}
H & H & H & O & H & H \\
| & | & | & \| & | & | \\
H-C-&C-&C-&C-&C-&C-H \\
| & | & | & | & | & | \\
H & H & H & H & H & H
\end{array}
$$

Ethyl propyl ketone

IUPAC naming techniques make use of the suffix *one* to indicate that we are dealing with a ketone. We identify the location of the doubly bound oxygen with the number of the carbon to which it is bound. Counting the number of carbons in the entire skeleton of ethyl propyl ketone, we realize that this compound is derived from hexane. The carbonyl carbon is the number 3 carbon because the chemical activity that produced the ketone is closest to the right-hand carbon. This ketone, by IUPAC procedures, is named 3-hexanone.

Example 16.3: Draw the compound that is composed of propyl and butyl functional groups that are joined by a carbonyl carbon. Name the compound.

Solution: We draw the compound that is described and find that it is a ketone.

$$
\begin{array}{ccccccccccccccc}
 & H & & H & & H & & O & & H & & H & & H & & H \\
 & | & & | & & | & & || & & | & & | & & | & & | \\
H- & C & - & C & - & C & - & C & - & C & - & C & - & C & - & C & -H \\
 & | & & | & & | & & & & | & & | & & | & & | \\
 & H & & H & & H & & & & H & & H & & H & & H
\end{array}
$$

This compound can be named propyl butyl ketone or 4-octanone, the IUPAC name.

Other Classes of
Organic Compounds

Example 16.4: What is the common name of the ketone that is composed of a propyl and a nonyl functional group joined by the carbonyl carbon?

Solution: Common names are derived from the functional groups that are joined by the carbonyl carbon. This compound is propyl nonyl ketone.

AMINES

The amine functional group is derived from ammonia, NH_3, by removing one hydrogen and bonding the amine group to a carbon. A primary amine is recognized by having only one of ammonia's hydrogens replaced with a hydrocarbon group, RNH_2. A secondary amine has two hydrogens replaced, R_2NH or $R'NHR''$. A tertiary amine has all three hydrogens replaced, R_3N.

$$H-N-R$$
$$|$$
$$H$$

A special amine-containing compound is an **AMIDE**. An amide is an organic acid substituted with an amine group. The amine group is found on the alpha carbon of amino acids; the alpha carbon is next to the carboxyl carbon. The importance of amino acids is that they are the building blocks of proteins. Amino acids form long chains by the amide of one amino acid bonding to the acid end of another during a dehydration process. This type of process is a **POLYMERIZATION**. *Polymers* are produced when individual repeating building blocks are put together. The result is the formation of a **PEPTIDE LINKAGE**, which is highly polar. The polarity of the peptide linkage, along with others and the disulfide linkages present, is responsible for the shape of the protein molecule. There are more than 20 amino acids commonly used to build proteins. Some of the R groups are polar.

Peptide Linkage

One of more important amines humans produce is urea. This amine is synthesized by the liver as a means for handling ammonia released during the deamination of proteins. Ammonia is extremely toxic, and two ammonias in the form of amides are tied up with a carbonyl group. Urea is toxic, but it is a great deal less so than is ammonia. Furthermore, urea is a molecule the kidneys are designed to handle; ammonia is not.

THIOLS

Thiols are much like alcohols. The thiol group contains a sulfur and the alcohol has an oxygen. Because oxygen and sulfur are so similar in chemical activity, a thiol functional group can be expected to behave in much the same manner as an alcohol group.

$$R-O-H$$
becomes
$$R-S-H$$

One of the more significant chemical reactions of thiols occurs when two cysteine molecules react. Cysteine is an amino acid and is present in many biologically important molecules, especially proteins. If two cysteine portions of a protein are close enough, they link (bond) to produce a **DISULFIDE LINKAGE**.

The disulfide linkage is very strong and is critical in determining the three-dimensional shape of the protein molecule in the location where the linkage forms. Enzymes are a class of proteins that have an active chemical function. The ability of the enzyme to work is in great part due to its shape.

AROMATICS

Aromatic compounds are generally considered to be based on ring (cyclic) structures with chemical bonding similar to benzene. The structure of benzene, C_6H_6, contains double bonds between alternating carbons, and the second bond is free to move. A chemical bond is a pair of electrons, so there are three pairs of electrons that are **DELOCALIZED** electrons. We indicate the structure of benzene with a hexagon with a circle in the center. The points on the hexagon are the locations of the carbons and their associated hydrogens. The circle represents the delocalized electrons.

The ability of the second bonds to move about the benzene molecule leads to **RESONANCE**. The concept of resonance notes that there are two ways to express the structure of benzene on the basis of where the second bonds are located over the passage of time. The following sketches of benzene assume that the carbon skeleton is held still in space and that we can look at the movement of the second bonds associated with the benzene ring.

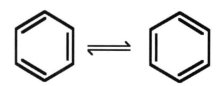

NOTE: There are no hydrogens directly indicated in the benzene structure. Recall that each carbon must have four chemical bonds. Since three of the four bonds are occupied with carbon-to-carbon bonding, the fourth bond is to a hydrogen.

193

Chapter 16
Other Classes of
Organic Compounds

We can substitute the benzene molecule, just as we substituted in previous sections. We replace a hydrogen (or more) with the substituent. The substitution can be performed with halides, functional groups, or even other benzenes. Incidentally, this substitution has no effect on resonance.

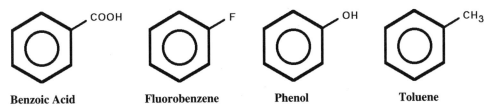

Benzoic Acid   **Fluorobenzene**   **Phenol**   **Toluene**

When we perform a single substitution on benzene, the carbon that receives the substituent becomes the number 1 carbon. The remainder of the carbons are numbered in sequence (2 through 6). (Phenol and toluene, organic solvents, are obviously not IUPAC names and must be classical names. Many of the aromatic compounds do not use IUPAC names.)

If we substitute with two or more substituents, the opportunity to produce isomers exists. If we use only two substituents, we have the choice of using the IUPAC numbering system for naming, or we can name according to the relative positions of the substituents.

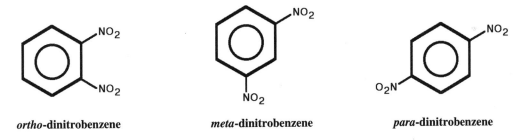

ortho-**dinitrobenzene**   *meta*-**dinitrobenzene**   *para*-**dinitrobenzene**

Abbreviations exist for ortho (*o*), meta (*m*), and para (*p*). The abbreviations are often used instead of the full word in this naming procedure. We can name these isomers as *o*-dinitrobenzene, *m*-dinitrobenzene, and *p*-dinitrobenzene.

The IUPAC naming system calls for the use of the carbon number in the name. We choose the number 1 carbon on the basis of where the most substitution has been performed, as we do with the substituted hydrocarbons. Once we have chosen the number 1 carbon, we count to the closest substituent and name that carbon. Then *o*-dinitrobenzene becomes 1,2-dinitrobenzene, *m*-dinitrobenzene becomes 1,3-dinitrobenzene, and *p*-dinitrobenzene becomes 1,4-dinitrobenzene. This same system lends itself to naming compounds with more than two substituents.

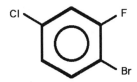

1-bromo 4-chloro 2-fluorobenzene

Quite a few biologically important compounds contain the benzene ring. Two examples are phenylalanine and tyrosine. These amino acids are found in many proteins synthesized by living organisms. The phenylalanine molecule is an alanine molecule substituted on the beta carbon by a benzyl group (the benzene ring). Tyrosine is an alanine molecule substituted on the beta carbon with a phenyl group (phenol less one hydrogen from the ring) in the para position.

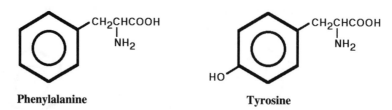

Phenylalanine   **Tyrosine**

There are compounds produced from benzene rings that share carbons. As with the other aromatic compounds, these compounds can be substituted. The simplest of the unsubstituted compounds is naphthalene, which is composed of two benzene rings. Mothballs are composed of naphthalene. Anthracene is produced from three benzene rings joined side by side. We can produce a compound of three benzene rings, two in a row and one above on the end. This isomer of anthracene is phenanthrene. The phenanthrene structure is one that you will see when you study cholesterol's structure, sex hormones, and other biologically important compounds.

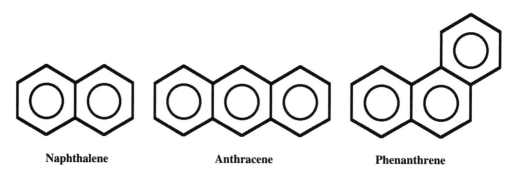

Naphthalene   **Anthracene**   **Phenanthrene**

Quite a few of these compounds are present in cigarette smoke, burning charcoal, wood smoke, and smoke from burning fossil fuels. An interesting point about these multiple benzene ring compounds (polycyclic aromatics) is that many of them are carcinogenic. A clue that a compound of this nature might be **CARCINO-GENIC** is obtained when we put a dot in the middle of the circles. We connect the dots with straight lines. If the line is bent, it is a clue that the compound is likely to be carcinogenic.

HETEROCYCLICS

The **HETEROCYCLIC** compounds contain at least one ring structure in which a minimum of one of the carbons has been replaced. The replacement is most often by one or more oxygen, nitrogen, or sulfur atoms. Many of these compounds are produced by living organisms. The pyridine ring is found in DPN and TPN, two hydrogen acceptors found in every living cell. The pyrimidine and purine rings serve as the skeleton for the nitrogen bases in DNA and RNA. The pyrimidines are cytosine (C), thymine (T), and uracil (U). The purines are adenine (A) and guanine (G).

Pyridine

Pyrimidine

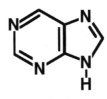

Purine

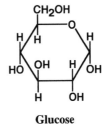

Glucose

Glucose is a heterocyclic compound in which there is one oxygen atom in the ring structure. We metabolize glucose to gain the energy that is stored in this molecule. The sources of glucose are the carbohydrates (glucose-containing sugars and starches) and sugars that can be converted into glucose (e.g. fructose).

There are isomers of glucose because the hydrogens and hydroxides can be found in different locations across the carbon to which they are bound.

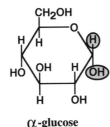

α-glucose

ß-glucose

In addition to the ring form, glucose opens to a straight-chain structure. We notice that the straight-chain form of glucose tells us that glucose is an aldehyde and a reducing sugar.

Glucose's aldehyde group can pick up a hydrogen, which is one of the ways of recognizing a reduction reaction. The reducing nature of a sugar can be detected by means of a chemical test, such as Benedict's solution. Benedict's solution turns from clear blue to cloudy brick red when reacting with a reducing sugar (the aldehyde group). The reason for the change in color is that Benedict's solution contains copper ions (Cu^{+2}). The copper ions are reduced by the aldehyde group, resulting in the production of a precipitate, copper I oxide.

$$
\begin{array}{c}
H \\
| \\
C=O \\
| \\
H-C-OH \\
| \\
HO-C-H \\
| \\
H-C-OH \\
| \\
H-C-OH \\
| \\
CH_2OH
\end{array}
$$

Fructose (fruit sugar) and glucose are isomers; they have the same molecular formula, $C_6H_{12}O_6$, but different structural formulas. The different structures change the nature of the sugars: Glucose is an aldehyde and fructose is a ketone when considered in the straight-chain representation. If you taste glucose and fructose, you find that fructose is a great deal sweeter than glucose. Fructose does not react with Benedict's solution or other solutions that test for reducing sugars. These are definitely not the same molecule.

$$
\begin{array}{c}
CH_2OH \\
| \\
C=O \\
| \\
HO-C-H \\
| \\
H-C-OH \\
| \\
H-C-OH \\
| \\
CH_2OH
\end{array}
$$

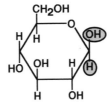

α-fructose

ß-fructose

Glucose and fructose are simple sugars. If we place two of these sugars next to each other, we have the opportunity to perform a dehydration reaction. The product of this reaction is a disaccharide. The bonding between the two sugar fragments is an important consideration because of the nature of the enzymes required. Generally speaking, enzymes are very specific regarding the reactions they catalyze. Alpha and beta glucose react by means of a dehydration, producing maltose and water.

α-Glucose   ß-Glucose   Maltose   Water

The linkage between the two glucose molecules in maltose is an α-1,4 linkage. The alpha linkage is recognizable as a u-shaped bond from the number 4 carbon of the right-hand glucose to the number 1 carbon of the left glucose.

Lactose has a ß-1,4 linkage between galactose's number 1 carbon and glucoses number 4 carbon. Galactose, glucose, and fructose are isomers ($C_6H_{12}O_6$).

ß-Galactose   α-Glucose   Lactose   Water

We notice that there is a significant difference between the alpha and beta linkage shown by maltose and lactose, respectively. Since enzymes tend to function on the basis of their structure, different enzymes must be required to make (and/or break) linkages that are so different in shape. We also see that the reactions producing maltose and lactose are reversible. Reactions that proceed in both directions are common in physiology.

Sucrose, table sugar, is also a disaccharide. Sucrose is the result of a dehydration reaction between glucose and fructose. Sucrose's linkage is an alpha linkage between the number 1 carbon of glucose and the number 2 carbon of fructose. Sucrose does not test

as a reducing sugar because the aldehyde functional group of sucrose is occupied with the linkage and fructose is not a reducing sugar.

Polymers of sugars are common in nature. Green plants produce glucose by the process of photosynthesis. Green plants have enzymes that can link glucose molecules to produce disaccharides such as maltose. However, there is no reason that more glucose molecules cannot be enzymatically added to the disaccharide to build a polysaccharide in steps. If this process is followed through, amylose, a starch, is produced.

Amylose is a straight chain of glucose molecules bound by alpha linkage from the number 4 carbon of one glucose to the number 1 carbon of another. The enzyme that performs this reaction is one that dehydrates (a dehydrase). We can digest this type of polysaccharide (a starch) by using an enzyme that can put the water back into the molecule. This a hydration reaction.

Plants also produce cellulose from the polymerization of glucose. Cellulose is very important to plants as a chemical structure that is found in the cell walls and is used as a stiffening material. Cellulose is produced by dehydration reactions, as is amylose. As with amylose, the linkage is from the number 4 carbon to the number 1 carbon of adjacent glucoses; however, the glucose molecules in cellulose are held together by beta bonds.

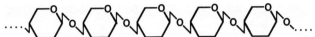

We cannot digest cellulose because we do not produce the appropriate enzyme that performs the hydrolysis of the beta bond. If humans could digest cellulose, plant materials would be a tremendous source of energy because a plant produces a lot of cellulose.

QUESTIONS AND PROBLEMS

1. What is the name of this compound?

$$
\begin{array}{c}
\ \ \ \ \text{H}\ \ \ \text{H}\ \ \ \text{H}\ \ \ \text{O} \\
\ \ \ \ |\ \ \ \ \ |\ \ \ \ \ |\ \ \ \ \ || \\
\text{H}-\text{C}-\text{C}-\text{C}-\text{C}-\text{H} \\
\ \ \ \ |\ \ \ \ \ |\ \ \ \ \ | \\
\ \ \ \ \text{H}\ \ \ \text{H}\ \ \ \text{H}
\end{array}
$$

2. Name this compound using the classical name and the IUPAC name.

$$
\begin{array}{c}
\ \ \ \ \text{O} \\
\ \ \ \ || \\
\text{H}-\text{C}-\text{H}
\end{array}
$$

3. Sketch the structural formula for pentyl aldehyde.

4. What is the structural formula for ethanal? Indicate the carbonyl carbon.

5. What is the name of this compound?

$$
\begin{array}{c}
\ \ \ \ \text{H}\ \ \ \text{H}\ \ \ \text{H}\ \ \ \text{O}\ \ \ \text{H}\ \ \ \text{H}\ \ \ \text{H} \\
\ \ \ \ |\ \ \ \ \ |\ \ \ \ \ |\ \ \ \ \ ||\ \ \ \ |\ \ \ \ \ |\ \ \ \ \ | \\
\text{H}-\text{C}-\text{C}-\text{C}-\text{C}-\text{C}-\text{C}-\text{C}-\text{H} \\
\ \ \ \ |\ \ \ \ \ |\ \ \ \ \ |\ \ \ \ \ \ \ \ \ |\ \ \ \ \ |\ \ \ \ \ | \\
\ \ \ \ \text{H}\ \ \ \text{H}\ \ \ \text{H}\ \ \ \ \ \ \ \text{H}\ \ \ \text{H}\ \ \ \text{H}
\end{array}
$$

6. Name this compound using the alkane functional group (alkyl) components and the IUPAC name.

$$
\begin{array}{c}
\ \ \ \ \text{H}\ \ \ \text{O}\ \ \ \text{H}\ \ \ \text{H}\ \ \ \text{H}\ \ \ \text{H}\ \ \ \text{H}\ \ \ \text{H} \\
\ \ \ \ |\ \ \ \ \ ||\ \ \ \ |\ \ \ \ \ |\ \ \ \ \ |\ \ \ \ \ |\ \ \ \ \ |\ \ \ \ \ | \\
\text{H}-\text{C}-\text{C}-\text{C}-\text{C}-\text{C}-\text{C}-\text{C}-\text{C}-\text{H} \\
\ \ \ \ |\ \ \ \ \ \ \ \ \ |\ \ \ \ \ |\ \ \ \ \ |\ \ \ \ \ |\ \ \ \ \ |\ \ \ \ \ | \\
\ \ \ \ \text{H}\ \ \ \ \ \ \ \text{H}\ \ \ \text{H}\ \ \ \text{H}\ \ \ \text{H}\ \ \ \text{H}\ \ \ \text{H}
\end{array}
$$

7.   Provide the structural formula of methyl propyl ketone and the IUPAC name.

8.   What are the locations of any polar points on methyl propyl ketone from problem 7?

9.   What is the difference between an amine and an amide?

10. The peptide linkages between the amino acids of a protein are partially responsible for the three-dimensional structure of the protein. What is the nature of the interaction between peptide linkages in this function?

11. Provide the formula for a thiol (the generalized formula) and indicate the place(s) that are likely to be polar.

12. Sulfur tends to be a negative polar point when found in a compound. Is the disulfide linkage polar?

13. Benzoic acid and phenol react to produce an ester by means of a dehydration reaction. Write the reaction using the structural formulas.

14—20: Name these compounds.

14. Br–⬡–Cl

15. Br–⬡ (Cl)

16. Br–⬡ (Cl) –Cl

17. (F, F, F on ring)

18. (F, O–H, F, F on ring)

19. Br–⬡–CH$_2$CH$_3$

20. (F, CH$_3$, F, F on ring)

21. What is the relationship of these two compounds to each other?

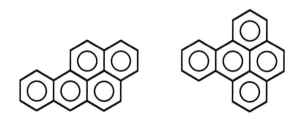

22. Page 195 contains the two isomeric structures of fructose. Page 196 presents the structures of two more sugars, glucose and galactose. What is the relationship of these carbohydrates?

23. What is the nature of the bonding between the hexose members of lactose? Comment on the digestability of this compound.

24. This disaccharide is sucrose. Can a polymer be produced from sucrose? If so, what is the nature of the chemical bonding?

25. This molecule is methylprednisolone, a glucocorticoid, which belongs to the compound class of the adrenocortical steroids. Methylprednisolone can be taken orally in pill form. Among its uses are the treatment of endocrine disorders, allergic reactions, respiratory disorders, and ophthalmic diseases.

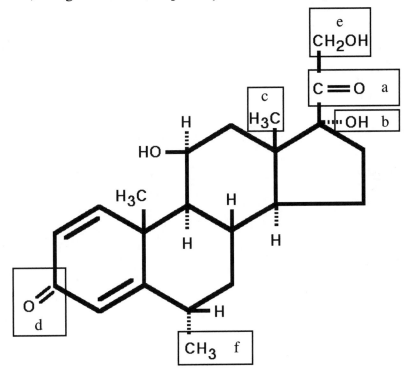

    (a)   What is the name of this functional group?
    (b)   What is the name of this functional group?
    (c)   What is the name of this functional group?
    (d)   What is the nature of this carbon?
    (e)   From what compound was this functional group derived?
    (f)   From what compound was this functional group derived?

SOLUTIONS AND ANSWERS

201

*Solutions
and
Answers*

Chapter 1

1—MEASUREMENT

1. (a) 1,000 mm   (b) 0.25 cm   (c) 320 m   (d) 0.0000005926 m
    (e) 0.015 L   (f) 0.000156 ML   (g) 6.2 μL   (h) 9 mL

2. ft + in = cm

$$\frac{in}{ft} \times ft + in = cm$$

$$\frac{cm}{in} \times \frac{in}{ft} \times ft + in = cm$$

$$\frac{2.54 \text{ cm}}{1 \text{ in}} \times \frac{12 \text{ in}}{1 \text{ ft}} \times 5 \text{ ft} + 6 \text{ in} = 167.64 \text{ cm}$$

The answer is most likely recorded as 167.6 cm.

3 . 0.3175 cm or 0.32 cm

4. (a) 998.22 cm; (b) 99.922 dm; (c) 9.98 m

5. (a) 487.68 cm; (b) 4.88 m (c) 0.0049 km

6. (a) 101.6 cm; (b) 0.102 m

7. (a) 1200 in^2; (b) 8.33 ft^2; (c) 0.926 yd^2; (d) 7741.9 cm^2; (e) 0.77 m^2

8. 103 cm^2

9. $$\frac{1^2 \text{ m}^2}{100^2 \text{ cm}^2} \times \frac{2.54^2 \text{ cm}^2}{1^2 \text{ in}^2} \times \frac{12^2 \text{ in}^2}{1^2 \text{ ft}^2} \times 11 \text{ ft} \times 16 \text{ ft} = 16.35 \text{ m}^2$$

10. 28.6 m^2, but the problem requires a whole number answer because carpet is usually purchased in a whole m^2 measure: 29 m^2.

11. $A = \pi r^2$ requires a radius, which is 1/2 of a diameter:
    $A = 3.1416 \times 2.25^2$ in = $3.14.16 \times 5.0625$ in = 15.90435 in^2
    Then
    $$A = 15.90435 \text{ in}^2 \times \frac{2.54^2 \text{ cm}^2}{1^2 \text{ in}^2} = 102.61 \text{ cm}^2$$

12. 2.68 in^2

13. 1 gallon will develop 100×8 in $\times 10$ in or 8,000 in^2. Each 5×7 is 35 in^2.
    Then 8,000 in^2/35 in^2 = 228.57 5×7's OR 228 pictures.

14. $$\frac{4 \text{ qt}}{1 \text{ gal}} \times \frac{1 \text{ L}}{1.06 \text{ qt}} \times \frac{\$0.35}{1 \text{ L}} = \frac{\$1.32}{\text{gal}}$$

15. 7.55 L

16. Since 1 L = 1.06 qt, the liter is the better deal by 6%.

17. $$\frac{1 \text{ L}}{1.06 \text{ qt}} \times \frac{1 \text{ qt}}{32 \text{ oz}} \times 14 \text{ oz} = 0.413 \text{ L}$$

18. 8.48 oz

19. 27.75 m^3

20. $$\frac{1 \text{ kg}}{2.2 \text{ lb}} \times 190 \text{ lb} = 86.4 \text{ kg}$$

21. (a) $\dfrac{1 \text{ kg}}{2.2 \text{ lb}} \times 190 \text{ lb} \times \dfrac{6 \text{ mg}}{\text{kg}} = 518.2 \text{ mg}$ (b) 1,140 mg

    (c) 1,140 mg - 518.2 mg = 621.8 mg overdose

22. (a) 1,818 kg;   (b) 1.82 Mg

23. 63.64 mL, but order 64 mL

24. $D = \dfrac{m}{V} = \dfrac{2{,}950 \text{ g}}{2 \text{ L} \times \dfrac{1{,}000 \text{ mL}}{1 \text{ L}}} = 1.475 \text{ g/mL}$

25. 0.539 g

26. 3,774 g or 3.8 kg

27. 319,980 g OR 704 lb

28. $\dfrac{2.2 \text{ lb}}{1 \text{ kg}} \times \dfrac{1 \text{ kg}}{1{,}000 \text{ g}} \times \dfrac{19.31 \text{ g}}{1 \text{ mL}} \times 5 \text{ gal} \times \dfrac{4 \text{ qt}}{1 \text{ gal}} \times \dfrac{1 \text{ L}}{1.06 \text{ qt}} \times \dfrac{1000 \text{ mL}}{1 \text{ L}} = 801.5 \text{ lb}$

    It would be nice to own all that gold, but I don't think so.

29. Lead is less dense than mercury; therefore, the lead will float on the mercury.

30. The density of water is assumed to be 100% of 1 g/mL and that of the ice is 1/9th of 1 g/mL based on the sunken portion of the iceberg.

    (1/9 = 0.89, so 0.89 × 1 g/mL = 0.89 g/mL)

    Then $SG = D_{ice}/D_{water} = 0.89$ [NOTE: no units]

31. 1.187

32. Hydrogen is less dense than air and will rise or, at least, reduce the weight of the container. A vacuum has no density; the hydrogen would not rise; the weight obtained would be a true measure.

33. $°F = 1.8°C + 32$ $°F - 32 = 1.8°C$ $\dfrac{°F - 32}{1.8} = °C$ $°C = \dfrac{98.6 - 32}{1.8} = 37.0$

34. 212°F

35. 77°F

36. $K = °C + 273$ $K = 25 + 273$ $K = 298$

37. 338.6K

38. 360.3

39. 0.004

40. Because the 12 has the lowest number of significant digits, the answer would be expressed in two significant digits.

2—ATOMIC STRUCTURE

1. The nucleus
2. The shells (orbits)
3. Protons
4. Since the atomic number is 2, there are two protons per atom.
5. 26 protons
6. Atomic number equals the number of protons. Since atoms have a charge of zero, the number of protons equals the number of electrons.
7. The atomic number of 92 is the number of protons = electrons.
8. 82 electrons

203

*Solutions
and
Answers*

Chapter 2
Chapter 3

9.   They don't. Neutrons have a zero charge.

10. 81 neutrons

11. 12 protons; 12 electrons; 12 neutrons

12. Neutrons have no charge. 11 protons contribute +11; 11 electrons have the charge of -11. The overall charge (+11 – 11) is zero.

13. Lithium has 3 protons; therefore, 3 electrons. The first orbit is filled with 2 electrons; the remainder flow into the next orbit. The answer is 2.

14. 4 electrons

15. 2 electrons

16. 8 electrons

17. 4 electrons; Group IVA elements have 4 electrons in the outside orbit.

18. 6 electrons by filling the orbits OR noticing that oxygen is in Group VIA.

19. Sulfur and oxygen are in Group VIA—they both have 6 electrons in the outside orbit.

20. The element with 79 protons (atomic number 79) is gold.

21. Californium (atomic number 98)

22. Both isotopes contain 17 protons and 17 electrons. Cl-35 has 18 neutrons and Cl-37 has 20 neutrons. The difference is the number of neutrons.

23. Since the two isotopes do contain the same number of electrons, the chemical reactivity should be approximately the same.

24. Chlorine-35 and chlorine-37

25. Since 12.011 is an AVERAGE atomic weight and very close to 12.000, the most common isotope by far is atomic weight 12.

26. 238

27. 44 amu

28. 98 amu

29. The total is 100% of W, which is 1 W. If one of the parts is W, then the other part is $1 - W$. To get the total

$$1W + (1 - W) = 1 \times 6.941$$

$$6(W) + 7(1 - W) = 1 \times 6.941$$

$$6W + 7 - 7W = 6.941$$

$$-1W + 7 = 6.941$$

Transpose to collect terms.

$$-1W = -0.059$$

Get rid of negative numbers on both sides. Multiply both sides by –1.

$$1W = 0.059$$

Since the other isotope is $1 - W$, then

$$1 - W = 1 - 0.059 = 0.94$$

One interpretation is that 94% of the lithium is Li-7 and 5.9% is Li-6.

A second interpretation is the ratio of Li-7:Li-6—divide either the decimal fractions or the percents: 94%/5.9% = 15.95, which means that there is approximately 16 times more Li-7 than Li-6 in existence.

30. 0.67% N-15

3—CHEMICAL COMPOUNDS

1.   Since Group VIIIA does not normally enter into chemical compounds, any element that has the same number of electrons in the orbits (especially the outside orbit) will tend

to be stable. *Stable* means that there will be no further gain or loss of electrons. Having 8 electrons in the outside orbit is stable. Having 2 electrons in the first orbit as the outside orbit is stable.

2. Metals tend to lose electrons; nonmetals tend to gain electrons.

3. The atomic number is 19; there are 19 protons and 19 electrons. We may leave out the neutrons since they have nothing to do with a stable electron configuration. Therefore, the portion of the structure we are interested in is

$$
\begin{array}{ccccc}
& \backslash & \backslash & \backslash & \backslash \\
19P & 2 & 8 & 8 & 1 \\
& / & / & / & /
\end{array}
$$

The nearest of the inert gases (Group VIIIA) are krypton (at. no. 36 with 4 orbits) and argon (at. no. 18 with 3 orbits). Since it would be much easier to look like argon than krypton (lose one electron rather than gain 7), potassium will lose one electron:

$$
\begin{array}{cccc}
& \backslash & \backslash & \backslash \\
19P & 2 & 8 & 8 \\
& / & / & /
\end{array}
$$

4.
$$
\begin{array}{ccc}
& \backslash & \backslash \\
13P & 2 & 8 \\
& / & /
\end{array}
$$

5. "The least amount of energy" means that there will be the least number of electrons moved; there will be 8 in the 3rd orbit due to gaining 2 electrons.

6. Carbon and silicon are in Group IVA; they will have 4 electrons in the outside orbit of their atoms. It is just about as easy to gain OR lose 4 electrons to reach stability. However, carbon is a nonmetal and will be predicted to prefer an electron gain. Silicon is a metalloid and will be predicted either to gain or lose electrons by choice. Also, since silicon is below carbon in Group IVA, the prediction will be that silicon tends to prefer an electron loss when compared to carbon's behavior.

7. Since radium is in Group IIA, there are 2 e⁻ in the outside orbit of the atom, the 7th orbit (Period 7). To become stable, radium will tend to lose 2 e⁻, rather than gain 6. The stable electron configuration will be just like radon (Rn, atomic number 86) with 8 electrons in the sixth orbit.

8.

Atom	Group	Outside orbit	e⁻ moved	Gain/Loss	Valence
Na	I	1	1	L	+1
Rb	I	1	1	L	+1
Mg	II	2	2	L	+2
Ca	II	2	2	L	+2
Sr	II	2	2	L	+2
B	III	3	3	L	+3
Ga	III	3	3	L	+3
Fr	I	1	1	L	+1

9. Sulfur is a member of Group VIA; therefore, the sulfur atom has 6 electrons in the outside orbit.

Sulfur can gain 2 electrons to produce the sulfide ion, S^{-2}.

Sulfur can lose 6 electrons to resemble neon; S^{+6}.

Sulfur can become an imitator of helium by losing 4 electrons; S^{+4}.

10. -1, +7, +5

11. +4, -4, +2

12. Iodide = I^{-1}; Sulfide = S^{-2}; Sulfate = SO_4^{-2}

13. Oxide = O^{-2}; Hydroxide = OH^{-1}; Potassium = K^{+1}; Phosphate = PO_4^{-3};
    Nitrate = NO_3^{-1}; Nitride = N^{-3}; Carbide = C^{-4}; Carbonate = CO_3^{-2};
    Bicarbonate = HCO_3^{-1}; Ammonium = NH_4^{+1}; Cyanide = CN^{-1}.

14. K^{+1} bonds with I^{-1}. Since the overall charge on a molecule is zero, KI ($+1-1 = 0$).
    Alternate solution: $K^{+1}I^{-1}$ is written as KI.

15. Since the overall charge on a molecule is zero, Na_2SO_4 ($+2-2 = 0$). Alternate solution:
    $Na^{+1}SO_4^{-2}$ is written as Na_2SO_4. Note that the parentheses are only required when more
    than one polyatomic ion is required.

16. Na is $+1$. Four oxygens total -8. Compounds have an overall charge of zero; therefore,
    the sulfur must be $+6$.

17. $+4$

18. Sodium cyanide = NaCN; Sodium carbonate = Na_2CO_3;
    Magnesium cyanide = $Mg(CN)_2$; Magnesium carbonate = $MgCO_3$;
    Magnesium phosphate = $Mg_3(PO_4)_2$; Aluminium phosphate = $AlPO_4$;
    Aluminium carbonate = $Al_2(CO_3)_3$; Aluminium cyanide = $Al(CN)_3$;
    Potassium sulfate = K_2SO_4; Beryllium nitrate = $Be(NO_3)_2$;
    Calcium permanganate = $Ca(MnO_4)_2$; Sodium bicarbonate = $NaHCO_3$;
    Boron hydroxide = $B(OH)_3$; Ammonium chloride = NH_4Cl;
    Aluminium sulfite = $Al_2(SO_3)_3$; Lithium nitrite = $LiNO_2$;
    Beryllium cyanide = $Be(CN)_2$; Ammonium phosphate = $(NH_4)_3PO_4$;
    Calcium carbonate = $CaCO_3$; Potassium chromate = K_2CrO_4.

19. Oxygen can be predicted to be a -2, $+6$, or $+4$ (Group VIA) in a compound. However,
    oxygen is a -2 unless it is bound to fluorine. Carbon could be a $+4$, -4, or $+2$ (Group
    IVA) in a compound. Carbon can only display the positive charges since oxygen can
    only be a -2. All compounds have a zero overall charge. Therefore, there are 2
    compounds: With carbon $+2$ ($+2-2 = 0$): CO. With carbon $+4$ ($+4-4 = 0$), but divide
    by the common factor of 2: CO_2.

20. Phosphorus and chlorine = PCl_3 and PCl_5; Nitrogen and oxygen = N_2O_3 and N_2O_5;
    Silicon and sulfur = SiS and SiS_2; Arsenic and nitrate = $As(NO_3)_3$ and $As(NO_3)_5$.

21. HNO_3 is nitric acid. The known charges are hydrogen $+1$ and oxygen -2 each. The total
    charge of the hydrogen is $+1$; the total charge of the oxygen is -6. By definition, the
    net charge on a compound is zero. The nitrogen must be $+5$ to satisfy the net charge
    of a compound.

22. BN	35. $(NH_4)_2S$	48. NH_4Cl	61. $HC_2H_3O_2$
23. $+3$	36. AsF_3	49. Li_2O	62. $SbCl_4$
24. $+4$	37. $Be_3(PO_4)_2$	50. $Al_2(CO_3)_3$	63. Ag_2S
25. CaS	38. $NaHCO_3$	51. BF_3	64. U_2S_5
26. SF_6	39. $Mg(HSO_4)_2$	52. Na_2S	65. SO_3
27. ClF_7	40. $BiCl_5$, $BiCl_3$	53. RaF_2	
28. $Cu(NO_3)_2$	41. $Al(OH)_3$	54. $Be(C_2H_3O_2)_2$	
29. Fe_2O_3	42. ZnO	55. $Mg(H_2PO_4)_2$	
30. Na_2O	43. $Ca(C_2H_3O_2)_2$	56. $Pb(OH)_4$	
31. K_3P	44. NaOH	57. $(NH_4)_3PO_4$	
32. LiI	45. HNO_3	58. $KHSO_4$	
33. $CaCO_3$	46. NaCN	59. Na_4C	
34. $B(C_2H_3O_2)_3$	47. K_2HPO_4	60. H_2SO_4	

1. The atomic weight of uranium is 238.0289 amu; one mole is 238.0289 grams.

2. 55.874 grams

3. $22.98977 \; ^g/_{mole} \times 18 \; \text{moles} = 413.82$ grams

4. 16.20 grams

5. The formula for calcium chloride is $CaCl_2$. The atomic weights from the Periodic Table are 40.078 for calcium and 35.4527 for chlorine. There are two chlorines in calcium chloride; the total weight from chlorine is 70.9054. Then the molecular weight of calcium chloride is $40.078 \text{ g Ca} + 70.9054 \text{ g Cl} = 110.9834 \; ^g/_{mol}$ calcium chloride.

6. $526.816 \; ^g/_{mol}$

7. The weight of one mole of H_2SO_4 is $98.0794 \; ^g/_{mol}$. The weight of three moles H_2SO_4 is $3 \; \text{mol} \times 98.0.794 \; ^g/_{mol} = 294.2382$ g.

8. HNO_3 weighs $63.01284 \; ^g/_{mol}$. Then $0.625 \; \text{mol} \times 63.01284 \; ^g/_{mol} = 39.383025 \text{ g} = 39.4$ g.

9. The formula for lithium iodide is LiI. Adding the atomic weights of one lithium and one iodine will give us the molecular weight of lithium iodide:
$$6.941 \; ^g/_{mole \; Li \; in \; LiI} + 126.9045 \; ^g/_{mole \; I \; in \; LiI} = 133.85 \text{ g LiI}$$

10. Magnesium sulfide is MgS; 24.3050 g Mg and 32.066 g S required.

11. The formula for carbonic acid is H_2CO_3.

$2 \times H$	+	$1 \times C$	+	$3 \times O$		
$2 \times 1.0079 \text{ g H}$	+	$1 \times 12.011 \text{ g C}$	+	$3 \times 15.9994 \text{ g O}$		
2.0158 g H	+	12.011 g C	+	47.9982 g O	=	$62.025 \; ^g/_{mole} \; H_2CO_3$

Therefore, 2.0158 grams hydrogen, 12.011 grams carbon, and 47.9982 grams oxygen are required to produce one mole H_2CO_3 ($62.025 \; ^g/_{mole}$).

12. Nitric acid is HNO_3. 1.0079 g hydrogen, 14.00674 g nitrogen, and 47.9982 g oxygen are required for each mole of nitric acid.

13. 75.26% Ag

14. 14.47% C

15. 20.01% S

16. We set up the two acids against each other and calculate the % S each.

$$\textit{for } H_2SO_3: \quad \frac{32.066}{2 \times 1.0079 + 32.066 + 3 \times 15.9994} \times 100 = 39.07\% \text{ S}$$

$$\textit{for } H_2SO_4: \quad \frac{32.066}{2 \times 1.0079 + 32.066 + 4 \times 15.9994} \times 100 = 32.69\% \text{ S}$$

Sulfurous acid has the greater amount of sulfur. The ratio of the sulfur content in sulfurous versus surfuric acids is 1.20:1.

17. The higer concentration of copper is from copper II carbonate.
    $Cu(NO_3)_2$ is 33.88 % Cu; $CuCO_3$ is 51.43% Cu.

18. Magnesium carbonate *appears* to be the better buy. $MgCO_3$ is 28.83% Mg and $MgCl_2$ is 25.53% Mg. However, we are given 1 kg of each compound, which is 11.86 moles $MgCO_3$ and 10.50 moles $MgCl_2$. We take the percent composition by moles of magnesium and each of the compounds and find that $MgCO_3$ *is* a better buy (3.4 moles Mg) than is $MgCl_2$ (2.68 moles Mg).

19. Assume there is a 100-g sample; there are 42.88 g C and 57.12 g O. Convert to moles.

$$42.88 \text{ g C} \times \frac{1 \text{ mol C}}{12.011 \text{ gC}} = 3.57006 \text{ mol C} \qquad 57.12 \text{ g O} \times \frac{1 \text{ mol O}}{15.9994 \text{ g O}} = 3.57001 \text{ mol O}$$

We have a 1:1 mole ratio of carbon to oxygen. The empirical formula is CO.

20. FeS

21. The formula for the compound is Si_xO_y. To provide the unknowns, x and y, we convert the grams of each into whole numbers of moles:

$$2.809 \text{ g Si} \times \frac{1 \text{ mol Si}}{28.0855 \text{ g Si}} = 0.1 \text{ mol Si} \quad and \quad 3.1999 \text{ g O} \times \frac{1 \text{ mol O}}{15.9994 \text{ g O}} = 0.2 \text{ mol O}$$

We cannot write $Si_{0.1}O_{0.2}$. Multiplication of both mole answers by 10 provides us with the whole number relationship: SiO_2.

22. We calculate the amount of carbon in the carbon dioxide sample and the amount of hydrogen in the water sample by working with the percents of carbon in the sample and hydrogen in the water:

for CO_2: $19.61 \text{ g } CO_2 \times \dfrac{12.011 \text{ g C}}{12.011 \text{ g C} + 2 \times 15.9994 \text{ g O}} = 5.351923 \text{ g C}$

for H_2O: $10.71 \text{ g } H_2O \times \dfrac{2 \times 1.0079 \text{ g H}}{2 \times 1.0079 \text{ g H} + 15.9994 \text{ g O}} \times 100 = 1.198389 \text{ g H}$

We change the gram values of each to moles in order to write the chemical formula:

$$5.35189 \text{ g C} \times \frac{1 \text{ mol C}}{12.011 \text{ g C}} = 0.44558 \text{ mol C}$$

$$1.198389 \text{ g H} \times \frac{1 \text{ mol H}}{1.0079 \text{ g H}} \times 100 = 1.188996 \text{ mol H}$$

We divide both of the values by the smallest in order to achieve a whole number relationship: 1 mol C:2.67 mol H.

We multiply by 3 because of the 0.67, which is the decimal equivalent of 2/3's; multiplication by 3 converts to a whole number. The empirical formula is C_3H_8.

23. C_2H_4

24. We calculate the amount of carbon and hydrogen by multiplying the collected products by the fraction of carbon/carbon dioxide and hydrogen/water.

Carbon from CO_2: $3.6162 \text{ g } CO_2 \times \dfrac{12.011 \text{ g C}}{44.0098 \text{ g } CO_2} = 0.9869 \text{ g C}$

Hydrogen from H_2O: $2.2204 \text{ g } H_2O \times \dfrac{2.0158 \text{ g H}}{18.0152 \text{ g } H_2O} = 0.2485 \text{ g H}$

We subtract the carbon and hydrogen from the sample weight and have the weight of the oxygen: 2.5500 g sample – 0.9869 g C – 0.2485 g O = 1.315 g O. We calculate the number of moles each. Divide through by the smallest to get a whole number ratio.

$0.9869 \text{ g C} \times \dfrac{1 \text{ mol C}}{12.011 \text{ g C}} = 0.08217 \text{ mol C}$ $0.08217 / 0.08217 = 1 \text{ carbon}$

$0.2485 \text{ g H} \times \dfrac{1 \text{ mol H}}{1.0079 \text{ g H}} = 0.2465 \text{ mol H}$ $0.2465 / 0.08217 = 3 \text{ hydrogen}$

$1.3146 \text{ g O} \times \dfrac{1 \text{ mol O}}{15.9994 \text{ g O}} = 0.08217 \text{ mol O}$ $0.08217 / 0.08217 = 1 \text{ oxygen}$

The empirical formula is CH_3O.

25. C_2H_3O

26. Ca_2C

27. We isolate the carbon and hydrogen: $47.06 \text{ g } CO_2 \times \dfrac{12.011 \text{ g C}}{44.0098 \text{ g } CO_2} = 12.84 \text{ g C}$

$$19.27 \text{ g } H_2O \times \dfrac{2.0158 \text{ g H}}{18.0152 \text{ g } H_2O} = 2.1562 \text{ g H}$$

We find the mole ratio.   We divide by the smallest number.

$12.84 \text{ g C} \times \dfrac{1 \text{ mol C}}{12.011 \text{ g C}} = 1.069 \text{ mol C}$ $1.069/1.069 = 1 \text{ carbon}$

$2.1562 \text{ g H} \times \dfrac{1 \text{ mol H}}{1.0079 \text{ g H}} = 2.139 \text{ mol H}$ $2.139/1.069 = 2 \text{ hydrogens}$

The empirical formula is CH_2. The empirical formula weight is 14.0268. Dividing the molecular weight by the empirical formula, 42.0804/14.0268, tells us we need 3 empirical formulas to determine the molecular formula, C_3H_6.

28. C_8H_{18}. This is the formula for octane, a component of gasoline.

29. The way to solve this problem is first to caluclate the amount of carbon and hydrogen from the combustion products.

$$3.81085 \text{ g } CO_2 \times \dfrac{12.011 \text{ g C}}{44.0098 \text{ g } CO_2} = 1.04 \text{ g C}$$

$$1.5601 \text{ g } H_2O \times \dfrac{2.0158 \text{ g H}}{18.0152 \text{ g } H_2O} = 0.17457 \text{ g H}$$

The second step is to subtract the carbon and hydrogen weights from the original sample weight to find the oxygen: 2.60 - 1.04 - 0.17457 = 1.385 g O.

The third step is to convert to moles of each component of the compound.

$$1.04 \text{ g C} \times \dfrac{1 \text{ mol C}}{12.011 \text{ g C}} = 0.8659 \text{ mol C}$$

$$0.17457 \text{ g H} \times \dfrac{1 \text{ mol H}}{1.0079 \text{ g H}} = 0.17320 \text{ mol H}$$

$$1.385 \text{ g O} \times \dfrac{1 \text{ mol O}}{15.9994 \text{ g O}} = 0.8657 \text{ mol O}$$

The next step is to obtain a whole number ratio for the empirical formula by dividing by the smallest number of moles.

0.8659 mol C/0.8657 = 1 mol C   0.17320 mol H/0.8657 = 2 mol H
0.8657 mol O/0.8657 = 1 mol oxygen   The empirical formula is CH_2O.

The last step is to divide the molecular weight by the empirical formula weight to determine how many empirical formulas are required for the molecular formula.
MW = ±120; EFW = ±32. 120/32 = 4 empirical formulas; $(CH_2O)_4$ is $C_4H_8O_4$.

30. C_2H_6O

31. C_4H_{10}

32. $C_{24}H_{60}Pb_3$

33. CuS

34. N_2O_4

35. U_2S_5; yes, inorganic compounds normally have a small, whole number relationship.

1. $2Li + S \rightarrow Li_2S$
2. $Mg + I_2 \rightarrow MgI_2$
3. $2B + 3Cl_2 \rightarrow 2BCl_3$
4. $2Zn + 3P \rightarrow Zn_3P_2$
5. $2Ba + C \rightarrow Ba_2C$
6. $2N_2 + 3O_2 \rightarrow 2N_2O_3$
7. $2Cu + Br_2 \rightarrow 2CuBr$
8. $2Fe + 3S \rightarrow Fe_2S_3$
9. $2S + 3O_2 \rightarrow 2SO_3$
10. $2Hg + Br_2 \rightarrow 2HgBr$
11. $4Al + 3O_2 \rightarrow 2Al_2O_3$
12. $Be + F_2 \rightarrow BeF_2$
13. $Si + O_2 \rightarrow SiO_2$
14. $Ni + S \rightarrow NiS$
15. $2Ca + C \rightarrow Ca_2C$
16. $2Ca + 2C + 3O_2 \rightarrow 2CaCO_3$
17. $H_2 + S + 2O_2 \rightarrow H_2SO_4$

18. Oxygen is higher in Group VIA than is sulfur and, therefore, will behave as a nonmetal with a negative oxidation number. Oxygen would tend to have an oxidation number of -2. The sulfur must be positive; there are two predictable oxidation numbers of +6 and +4. Therefore, there are two equations:
$$S + O_2 \rightarrow SO_2 \quad and \quad 2S + 3O_2 \rightarrow 2SO_3$$

19. $2N_2 + 3O_2 \rightarrow 2N_2O_3 \quad and \quad 2N_2 + 5O_2 \rightarrow 2N_2O_5$
20. $2Na + 2HCl \rightarrow 2NaCl + H_2$
21. $Mg + H_2SO_4 \rightarrow MgSO_4 + H_2$
22. $2Al + 3H_2SO_4 \rightarrow Al_2(SO_4)_3 + 3H_2$
23. $3Ra + 2H_3PO_4 \rightarrow Ra_3(PO_4)_2 + 3H_2$
24. $4Na + CCl_4 \rightarrow 4NaCl + C$
25. $3Ca + 2AlBr_3 \rightarrow 3CaBr_2 + 2Al$
26. $6K + SO_3 \rightarrow 3K_2O + S$
27. $2Fe_2S_3 + 3O_2 \rightarrow 2Fe_2O_3 + 6S$
28. $3Na + BPO_4 \rightarrow Na_3PO_4 + B$
29. $2Fr + CuCO_3 \rightarrow Fr_2CO_3 + Cu$
30. $Ag_2S + F_2 \rightarrow 2AgF + S$
31. No reaction. Fluorine is more active chemically than oxygen.
32. $Sr + CaCO_3 \rightarrow SrCO_3 + Ca$
33. $2Fe + 3H_2SO_4 \rightarrow Fe_2(SO_4)_3 + 3H_2$
34. $Cu + 2HC_2H_3O_2 \rightarrow Cu(C_2H_3O_2)_2 + H_2$
35. Yes. $2RaO + 2F_2 \rightarrow 2RaF_2 + O_2$
36. No reaction. Neon is an inert gas and is not active enough to replace either sodium or chlorine.
37. $H_2SO_4 + Be(OH)_2 \rightarrow BeSO_4 + 2HOH$
38. $3HCl + Al(OH)_3 \rightarrow AlCl_3 + 3HOH$
39. $2H_3PO_4 + 3Zn(OH)_2 \rightarrow Zn_3(PO_4)_2 + 6HOH$
40. $2HC_2H_3O_2 + Cu(OH)_2 \rightarrow Cu(C_2H_3O_2)_2 + 2HOH$
41. $5H_2CO_3 + 2U(OH)_5 \rightarrow U_2(CO_3)_5 + 10HOH$
42. $2AgNO_3 + MgCl_2 \rightarrow 2AgCl + Mg(NO_3)_2$
43. $2HCl + CaCO_3 \rightarrow CaCl_2 + CO_2 + H_2O$
44. $3Pb(NO_3)_2 + Al_2(SO_4)_3 \rightarrow 3PbSO_4 + 2Al(NO_3)_3$
45. $MgCl_2 + 2NaOH \rightarrow Mg(OH)_2 + 2NaCl$
46.
$$Zn \rightarrow Zn^{+2} + \cancel{2e^-}$$
$$\underline{Cu^{+2} + \cancel{2e^-} \rightarrow Cu}$$
$$Zn + Cu^{+2} \rightarrow Zn^{+2} + Cu$$

210

*Solutions
and
Answers*

Chapter 5
Chapter 6

47. $Ni^{+2} + Cd \rightarrow Ni + Cd^{+2}$

48. $\quad\quad Fe \rightarrow Fe^{+3} + 3e^- \quad\quad\quad Fe \rightarrow Fe^{+3} + \cancel{3e^-}$

$3[1e^- + Ag^{+1} \rightarrow Ag] \quad\quad \dfrac{\cancel{3e^-} + 3Ag^{+1} \rightarrow 3Ag}{Fe + 3Ag^{+1} \rightarrow Fe^{+3} + 3Ag}$

49. $2Al + 3Mg^{+2} \rightarrow 2Al^{+3} + 3Mg$

50. $2Cu + Ca^{+2} \rightarrow 2Cu^{+1} + Ca$

51. $3Ba + 2As^{+3} \rightarrow 3Ba^{+2} + 2As$

52. $3Sr + 2Al^{+3} \rightarrow 3Sr^{+2} + 2Al$

53. $2[4H^{+1} + 3e^- + NO_3^{-1} \rightarrow NO + 2H_2O]$

$\dfrac{3[S^{-2} \rightarrow S + 2e^-]}{8H^{+1} + 2NO_3^{-1} + 3S^{-2} \rightarrow 2NO + 3S + 4H_2O}$

54. $16H^{+1} + 2MnO_4^{-1} + 10Cl^{-1} \rightarrow 2Mn^{+2} + 5Cl_2 + 8H_2O$

55. The mercury is a spectator ion and is omitted. $S^{-2} + I_2 \rightarrow S + 2I^{-1}$

56. $4H^{+1} + Pb + 4NO_3^{-1} \rightarrow PbO_2 + 4NO_2 + 2H_2O$

57. $H_2O + Se + 2SnO_2^{-2} \rightarrow SeO_3^{-2} + 2Sn + 2OH^{-1}$

58. $8OH^{-1} + 5Ba^{+2} + 2N^{-2} \rightarrow 5Ba + 2NO_2^{-1} + 4H_2O$

59. $H_2O + Se + 2ZnO_2^{-2} \rightarrow SeO_3^{-2} + 2Zn + 2OH^{-1}$

60. $9OH^{-1} + 7S + 2Re \rightarrow 7HS^{-1} + 2ReO_4^{-1} + H_2O$

61. zinc	62. strontium	63. sulfur	64. iron
65. copper	66. aluminum	67. nitrogen	68. manganese
69. copper	70. aluminum	71. nitrogen	72. manganese
73. zinc	74. strontium	75. sulfur	76. iron

6—CALCULATIONS AND CHEMICAL REACTIONS

1. $\quad$ 2 mol $\quad$ 1 mol

$\quad$ $2Ag + S \rightarrow Ag_2S$ $\quad\quad \dfrac{2\text{ mol } \cancel{Ag}}{7\text{ mol } \cancel{Ag}} = \dfrac{1\text{ mol S}}{y}$ $\quad\quad 2y = 1$ mol S $\times 7$

$\quad$ 7 mol $\quad$ y $\quad\quad\quad\quad\quad\quad\quad\quad\quad\quad\quad\quad\quad\quad y = (1$ mol S $\times 7)/2$

$\quad\quad\quad\quad\quad\quad\quad\quad\quad\quad\quad\quad\quad\quad\quad\quad\quad\quad\quad\quad y = 3.5$ mol S

2. 30.67 mol Fe

3. 4 mol $HC_2H_3O_2$

4. 1.5 mol $Ra(OH)_2$

5. 8.67 mol H_3PO_4

6. $\quad\quad\quad\quad$ 1 mol $\quad\quad$ 1 mol

$\quad$ $2Na + H_2SO_4 \rightarrow Na_2SO_4 + H_2$

$\quad\quad\quad\quad\quad$ y $\quad\quad\quad$ 5 mol

$\quad \dfrac{1\text{ mol }H_2SO_4}{y} = \dfrac{1\text{ mol }Na_2SO_4}{5\text{ mol }Na_2SO_4}$ $\quad\quad y = 5$ mol H_2SO_4

7. 100 mol Na

8. 0.10 mol ozone

9. $\quad$ 4 mol $\times$ 63.456 $^g/_{mol}$ $\quad$ 31.9988 g O_2

$\quad\quad\quad$ $4Cu$ $\quad\quad\quad + \quad\quad O_2 \quad\quad \rightarrow 2Cu_2O$

$\quad\quad\quad\quad$ y $\quad\quad\quad\quad$ 3 g

$\quad \dfrac{4\text{ mol } \cancel{Cu} \times 63.546 \text{ }^{g\,Cu}/_{\cancel{mol\,Cu}}}{y} = \dfrac{31.9988\text{ g} \cancel{O_2}}{3\text{ g} \cancel{O_2}}$

$\quad y = \dfrac{4 \times 63.546\text{ g Cu} \times 3}{31.9988} = 23.8$ g Cu

10. 8.60 g Zn
11. 39.93 g S
12. 0.171 g Cr
13. 21.6 g HCN
14. 0.24 g H_2SO_4
15. 0.299 g Mg
16.

63.546   95.612 g

$$Cu + S \rightarrow CuS$$

y   2.9 g

$$\frac{63.546 \text{ g Cu}}{y} = \frac{95.612 \text{ g CuS}}{2.9 \text{ g CuS}} \qquad y = \frac{2.9 \text{ g CuS} \times 63.546 \text{ g Cu}}{95.612 \text{ g CuS}} = 1.9 \text{ g Cu}$$

17. 0.0035 g Th
18. 0.031g O_2
19. Let's use whole number atomic weights since the weights are given to us in whole numbers.

98 g   105 g

$$H_2SO_4 + Be(OH)_2 \rightarrow BeSO_4 + 2HOH$$

y   21 g

$$\frac{98 \text{ g } H_2SO_4}{y} = \frac{105 \text{ g } Be(OH)_2}{21 \text{ g } Be(OH)_2} \qquad y = \frac{21 \times 98 \text{ g } H_2SO_4}{105} = 20 \text{ g } H_2SO_4$$

20. 95.6 g $B(C_2H_3O_2)_3$
21. 83.13 g PCl_3
22. 120.234 g   61.94752 g

$$3Ca + 2P \rightarrow Ca_3P_2$$

0.155 g   0.332 g   y

$$\frac{120.234 \text{ g Ca}}{t} = \frac{61.94752 \text{ g P}}{0.332 \text{ g P}}$$

Since we have weights of both reactants, we test to see if there is a limiting reactant using t for the tested reactant.

$$t = \frac{0.332 \times 120.234 \text{ g Ca}}{61.94752} = 0.644 \text{ g Ca}$$

120.234 g   61.94752

$$3Ca + 2P \rightarrow Ca_3P_2$$

t   0.332

$t = 0.644$ g Ca required to use the phosphorous; we only have 0.155 g. Since calcium is the limiting reactant, we set up the portion of the equation that applies.

120.234 g   182.18152 g

$$3 \text{ Ca} \approx Ca_3P_2$$

0.155 g   y

$$\frac{120.234 \text{ g Ca}}{0.155 \text{ g Ca}} = \frac{182.18152 \text{ g } Ca_3P_2}{y}$$

$$y = \frac{182.18152 \text{ g } Ca_3P_2 \times 0.155}{120.234}$$

$y = 0.235$ g Ca_3P_2 predicted production.

23. 11.2 g ZnS   24. 16.97 g Na_2S 25. 59.7 g NO_2 26. 0.0362 g $(NH_4)_3PO_4$
27. 127 g AgCl   28. 18.09 g Fe   29. 4.47 g $BaSO_4$ 30. 53.74 g SO_3

7—GAS LAWS

1. The intermolecular forces that would hold the molecules in a crystalline structure cannot be established over the large distance between molecules.

2. Pressure and volume do not have to be converted. The conversions are muliplications by a factor; that factor appears on both sides of the equation and cancels. *REMINDER: Temperature conversions have an algebraic addition that cannot be canceled.*

3. Boyle's—Temperature   Charles'—Pressure   Gay-Lussac's—Volume

4. Avogadro's principle: The number of molecules is equal for equal volumes under the same conditions.

5. Dalton's law of partial pressure: The final pressure would be a simple addition of the pressures.

6. $PV = nRT$

$$1\ \text{atm} \times V = \frac{100\ \cancel{g}}{32\ \cancel{g}/\cancel{mol}} \times 0.0821\ \frac{\text{L atm}}{\cancel{mol}\,\cancel{K}} \times 298\cancel{K}$$

$$V = \frac{\dfrac{100}{32} \times 0.0821\ \text{L}\ \cancel{atm} \times 298}{1\ \cancel{atm}} = 76.5\ \text{L}$$

7. 11.2 atm

8. No, the container would not hold the pressure of 56.4 atmospheres

9. $\dfrac{P_1V_1}{T_1} = \dfrac{P_2V_2}{T_2}$

Since $T_1 = T_2$, they cancel. The *conversion* of torr to atm is a multiplication on both sides and cancels, torr may be used as is and does not need to be converted. *REMINDERS: If temperature were required, we would convert to Kelvin since the conversion does not cancel (addition). Also, one atmosphere of pressure is 760 torr.*

$$760\ \text{torr} \times 5\ \text{L} = 950\ \text{torr} \times V_2 \qquad V_2 = \frac{760\ \text{torr} \times 5\ \text{L}}{950\ \text{torr}} = 4\ \text{L}$$

10. 45 L

11. 15 L

12. $\dfrac{P_1V_1}{T_1} = \dfrac{P_2V_2}{T_2} \qquad \dfrac{0.9\ \text{atm} \times 2\ \text{L}}{303\text{K}} = \dfrac{3\ \text{atm} \times 0.5\ \text{L}}{T_2}$

$$0.9\ \text{atm} \times 2\ \text{L} \times T_2 = 303\text{K} \times 3\ \text{atm} \times 0.5\ \text{L}$$

$$T_2 = \frac{303\text{K} \times 3\ \text{atm} \times 0.5\ \text{L}}{0.9\ \text{atm} \times 2\ \text{L}} = 252.5\text{K}$$

The problem requests the answer in °C. Since K = °C + 273, then °C = K – 273. The substitution and algebraic addition provide us with -20.5°C.

13. 601.4K or 328.4C

14. By Dalton's law of partial pressure, the gases are added together to yield the total pressure, 2.5 atm (100%). The paritial pressure of oxygen is 35% of 2.5 atm. The partial pressure of nitrogen is 65% of 2.5 atm.

    Oxygen $0.35 \times 2.5\ \text{atm} = 0.875\ \text{atm}$
    Nitrogen $0.65 \times 2.5\ \text{atm} = 1.625\ \text{atm}$

15. Argon: 162 torr
    Krypton: 288 torr
    Xenon: 450 torr

16. $PV = nRT$ and $n = g/MW$. *REMINDER: STP is 1 atm and 0°C.*

$$PV = \frac{g}{MW} \times RT$$

$$1 \text{ atm} \times 0.25 \text{ L} = \frac{0.300 \text{ g}}{MW} \times 0.0821 \frac{L \text{ atm}}{\text{mol K}} \times 273K$$

$$1 \text{ atm} \times 0.25 \text{ L} \times MW = 0.300 \text{ g} \times 0.0821 \frac{L \text{ atm}}{\text{mol K}} \times 273K$$

$$MW = \frac{0.300 \text{ g} \times 0.0821 \frac{L \text{ atm}}{\text{mol K}} \times 273K}{1 \text{ atm} \times 0.25 L} = 26.9 \text{ g/mol}$$

17. 16 grams/mole
18. 57.88K
19. 0.26 atm
20. C = 42.88 g × 1 mole/12.011 g = 3.57 mole ~ 1 carbon
    O = 57.12 g × 1 mole/15.9994 g = 3.57 mole ~ 1 oxygen
    The mole ratio is 1:1; the empirical (simplest) formula is CO ($28^{g}/_{mole}$).

$$PV = nRT$$

$$1 \text{ atm} \times 40 \text{ L} = 100 \frac{g}{MW} \times 0.0821 \frac{L \text{ atm}}{\text{mol K}} \times 273K$$

$$MW \times 1 \text{ atm} \times 40 \text{ L} = 100 \text{ g} \times 0.0821 \frac{L \text{ atm}}{\text{mol K}} \times 273K$$

$$MW = \frac{100 \text{ g} \times 0.0821 \frac{L \text{ atm}}{\text{mol K}} \times 273K}{1 \text{ atm} \times 40 L} = 56.03 \frac{g}{\text{mol}}$$

The weight of CO is approximately 28 $^{g}/_{mol}$ and the MW = 56 $^{g}/_{mol}$; 2 EFs equal the MW. The molecular formula of the compound is C_2O_2.

21. We set up the Graham's law relationship and solve for the molecular weight.

$$\frac{R_1}{R_2} = \sqrt{\frac{MW_2}{MW_1}} \qquad \frac{1.33}{1} = \sqrt{\frac{MW_2}{17 \text{ g/mol}}} \quad \text{sq. both sides} \quad \frac{1.7689}{1} = \frac{MW_2}{17 \text{ g/mol}}$$

$$1 \times MW_2 = 1.7689 \times 17 \, ^{g}/_{mol} \qquad MW_2 = 1.7689 \times 17 \, ^{g}/_{mol} = 30.1 \, ^{g}/_{mol}$$

22. The relationship calls for a rate, not a time comparison. We divide the times to get the comparative rate: Dividing 35/52 = 0.673 sets the 35-minute data as "1" information and the 52-minute data as "2" information relative to rates of diffusion.

$$\frac{R_1}{R_2} = \sqrt{\frac{MW_2}{MW_1}} \qquad \frac{0.673}{1} = \sqrt{\frac{MW_2}{235 \text{ g/mol}}} \qquad \frac{0.673^2}{1} = \frac{MW_2}{235 \text{ g/mol}}$$

$MW_2 = 0.673^2 \times 235 \, ^{g}/_{mol} = 106.4 \, ^{g}/_{mol}$.
MW is in $^{g}/_{mol}$—the correct units.

23. We have to calculate the empirical formula to get to the molecular formula. We assume that we have 100 g of the hydrocarbon and will make use of the parts per hundred (%). In the case of the carbon we will have 92.258 g to work with—we need to express that in moles because the chemical formula will be a mole carbon to mole hydrogen ratio (and, of course, do the same thing with the hydrogen):

$$92.258 \text{ g C} \times \frac{1 \text{ mol C}}{12.011 \text{ g C}} = 7.6811 \text{ mol C} \qquad 7.742 \text{ g H} \times \frac{1 \text{ mol H}}{1.0079 \text{ g H}} = 7.6813 \text{ g H}$$

214

**Solutions
and
Answers**

Chapter 7
Chapter 8

The mole ratio is a 1:1 ratio; the empirical formula is CH. Next, we find the molecular weight so we know how many empirical formulas we will need for the molecular formula.

$$\frac{R_{ammonia}}{R_{hydrocarbon}} = \sqrt{\frac{MW_{hydrocarbon}}{MW_{ammonia}}} \qquad \frac{1.236}{1} = \sqrt{\frac{MW_{hydrocarbon}}{17.03044 \ g/mol}}$$

We square both sides to eliminate the sqaure roots.

$$\frac{1.236^2}{1} = \frac{MW_{hydrocarbon}}{17.03044 \ g/mol}$$

$MW_{hydrocarbon} = 1.236^2 \times 17.03044 \ g/mol = 26.017335 \ g/mol$

Since CH weighs $13^g/_{mole}$, the molecular formula of the hydrocarbon is C_2H_2 (acetylene).

24. The molecular weights are 36.15 for HCl and 159.8 for bromine (Br_2). HCl has a lower molecular weight and, therefore, diffuses 2.1 times faster.

25. The MW of 213 is a trap. We don't need MWs since the relative speeds were given, as was the speed of one of the gases. We can just set up a ratio and proportion of the faster to the slower.

$$\frac{4.5}{1} \bigtimes \frac{4.0 \ \frac{cm}{min}}{y} \qquad 4.5 \times y = 4.0 \ \frac{cm}{min} \qquad y = \frac{4.0 \ \frac{cm}{min}}{4.5} = 0.89 \ \frac{cm}{min}$$

The rate of the slower is 0.89 cm/min.

26. Information required: original temperature, volume, and pressure—the temperature of the environment (ambient temperature) and the barometric pressure.

Equation required: combined gas law equation since there are two sets of conditions.

27. Graham's law is stated in MW, which is grams/mol of gases. Density is worked out in grams/liter. According to Avogadro's principle, gases take up a specific volume under specific conditions. From Graham's law we use moles of a gas and, from Avogadro's principle, we can covert to liters; grams/mole (MW) can, therefore, be expressed in grams/liter (density). Since all of the conversions are multiplied AND are on opposite sides of an equation made up of two fractions equal to each other AND the conversions are in the same location AND multiplied/divided, they cancel.

28. Large molecules are the best choice. The larger the molecule, the slower will be the diffusion and the longer the scent will last.

29. The effect of an increase in temperature on molecules of a particular gas is an increase in the velocity of the molecules. The rate of diffusion is dependent on the velocity of the molecules of a particular gas. Therefore, the gas diffuses faster at a higher temperature than at a lower one.

30. The dry gas from the heated test tube passes through the water. Water vapor is picked up by the bubbles of gas. The collection test tube contains both the gas produced and the water vapor from the water that evaporates into the bubbles of gas.

8—CHEMICAL EQUILIBRIA

1. $K_c = \dfrac{[SO_3][NO]}{[SO_2][NO_2]}$ 2. $K_c = \dfrac{[SO_3]^2}{[SO_2]^2[O_2]}$ 3. $K_c = \dfrac{[CH_4][H_2S]^2}{[H_2]^4[CS_2]}$

4. $K_c = \dfrac{[NO]^4[H_2O]^6}{[NH_3]^4[O_2]^5}$

5. $K_c = \dfrac{[CO][Cl_2]}{[COCl_2]}$

6. $K_c = \dfrac{[NO_2]^2}{[NO]^2[O_2]}$

7. $K_c = \dfrac{[CO_2][H_2]}{[CO][H_2O]}$

8. $K_c = \dfrac{[N_2O_4]}{[NO_2]^2}$

9. $K_c = \dfrac{[NH_3]^2}{[H_2]^2[N_2]^3}$

10. $K_c = \dfrac{[H_2][I_2]}{[HI]^2}$

11. $K_c = \dfrac{\text{product of products}}{\text{product of reactants}} = \dfrac{[CO][Cl_2]}{[COCl_2]} = \dfrac{(0.022/1)(0.022/1)}{0.50/1} = 9.7 \times 10^{-4}$

12. $K_c = 2.6$

13. $K_c = 5.9 \times 10^{-2}$

14. 13.1 mol/L A

15. We set up a table so we see what happens during the reaction and what the final concentrations are for calculation purposes.

	$COCl_2 \rightleftharpoons$	CO	$+ Cl_2$
Original	1.9/1	0	0
Gain/Loss	$-y$	$+y$	$+y$
Final	$1.9 - y$	y	y

We set up the K_c formula, substitute the final concentrations, and solve.

$K_c = \dfrac{[CO][Cl_2]}{[COCl_2]}$ $0.575 = \dfrac{y^2}{1.9 - y}$ $y^2 = 0.575(1.9 - y)$
$y^2 = 1.0925 - 0.575y$

The solution to this problem requires the use of the quadratic formula. We rearrange the equation to $y^2 + 0.575y - 1.0925 = 0$ and substitute into the quadratic formula. We receive only one answer that makes sense, $y = 0.7965462$; the other answer is a negative number. The value of y is the concentration of *both* CO and Cl_2, according to the table we set up previously. Taking into account significant digits, $y = 0.80$ mol CO/L.

16. $K_c = \dfrac{[SO_3]^2}{[SO_2]^2[O_2]}$ $K_c = \dfrac{(2.75/8)^2}{(4.00/8)^2(4.00/8)}$ $K_c = 0.945$

17. $K_c = 58$

18. $K_c = 5.121$

19. $K_c = 88.9$

20.

	$H_2 +$	$I_2 \rightleftharpoons$	$2HI$
Original	0	0	2
Gain/Loss	$+y$	$+y$	$-y$
Final	y	y	$2-y$

$K_c = \dfrac{[HI]^2}{[H_2][I_2]}$ $45.0 = \dfrac{(2-y)^2}{(y)(y)}$

Take the square root of the equation, cross multiply, and solve.

$6.7082039 = \dfrac{2-y}{y}$ $6.7082y = 2 - y$ $7.7092039y = 2$ $y = 0.2594638$ $y = 0.26$

We do not have an answer to this problem quite yet. We finish this problem by substituing the value of y for the final concentrations.

Final y y $2-y$
                0.26   0.26 $2-0.26 = 1.74$
                H_2 I_2 HI concentrations in mol/L (molarity).

21. NOTE: K'_c is the inverse of K_c. $K'_c = \dfrac{[N_2O_4]}{[NO_2]^2}$

22. $K'_c = \dfrac{[HI]^2}{[H_2][I_2]}$

23. $K'_c = 0.6154$

24. The K'_c is the inverse of the K_c; therefore, $1/33.33 = 0.03$.

25. The K_c does not include solids and liquids because they are of a constant volume.
     Therefore, $K_c = [CO_2]$.

26. $K'_c = \dfrac{[CO]^2}{[CO_2]}$

27. $K_c = [Br_2]$

28. $K_c = [SO_3]$

29. The reaction we use to calculate K'_c is the decomposition reaction.
     $NH_4Cl_{(s)} \rightleftharpoons NH_{3\,(g)} + HCl_{(g)}$ $K'_c = [NH_3][HCl] = (^{0.15}/_3)^2 = 2.5 \times 10^{-3}$

30. $K_c = 11.75 = \dfrac{[H_2O]}{[H_2]} = \dfrac{y/\cancel{4}}{2/\cancel{4}}$ Therefore, $11.75 = \dfrac{y}{2}$

     Cross multiplication simplifies the equation: $y = 2 \times 11.75 = 23.50$ mol $H_2O_{(g)}$.

31. *NOTE: This problem gives us the volume of the container. We do not need this
     information since we solve on the basis of pressures (atmospheres).*

     $K_p = \dfrac{(P_{SO_3})(P_{NO})}{(P_{SO_2})(P_{NO_2})}$ $K_p = \dfrac{0.600 \times 0.400}{1.500 \times 1.500} = 0.107$

32. $K_p = 0.025$

33. $K_p = 0.0625$

34. We are required to work K_p problems in terms of atmospheres. We convert from grams
     to atmospheres by $PV = nRT$, which can be rearranged to

$$P_{NO_2} = \frac{^{10.5}\cancel{g}/46\ \text{g/mol} \times 0.0821\ ^{L\ atm}/_{mol\ K} \times 593\ K}{2\ L} = 5.56\ \text{atm}$$

$$P_{N_2O_4} = \frac{^{35}\cancel{g}/92\ \text{g/mol} \times 0.0821\ \cancel{L}\ atm/_{\cancel{mol}\ \cancel{K}} \times 593\ \cancel{K}}{\cancel{2\ L}} = 9.26\ \text{atm}$$

$$K_p = \frac{(P_{N_2O_4})}{(P_{NO_2})^2} = \frac{9.26}{5.56^2} = \frac{9.26}{30.91} = 0.30$$

35. *NOTE: This is a heterogeneous reaction; pure solids and liquids are not included in
     the calculation of K_p.* $K_p = 0.33$.

1. Percent (wt/wt) is calculated by the weight of the solute divided by the weight of the entire solution and then multiplied by 100.

$$\text{percent solution} = \frac{\text{grams solute}}{\text{g solute} + \text{g solvent}} \times 100$$

$$\text{percent solution} = \frac{10 \text{ g NaCl}}{10 \text{ g NaCl} + 90 \text{ g H}_2\text{O}} \times 100 = 90\%$$

This solution is 90% NaCl (wt/wt).

2. 25.9% H_2SO_4

3. 0.826% tincture of iodine

4. We calculate the mass of the unknown solute by setting up a ratio and proportion in which the known information is set equal to the solution to be mixed.

$$\frac{25 \text{ g solute}}{100 \text{ g solution}} = \frac{y}{200 \text{ g solution}}$$

$$y = \frac{25 \text{ g solute} \times 200 \text{ g solution}}{100 \text{ g solution}} = 50 \text{ g solute}$$

Since we are to mix 200 g of solution and know we are required to use 50 g solute, we need 150 g water (200 – 50 = 150).

5. 150 g NaCl and 850 g H_2O

6. A wt to vol solution is relative to 100 mL of solution. We are mixing a solid with water and must assume that the final volume of the solution is 250 mL. We set up a ratio and proportion, which provides us with the percent composition.

$$\frac{25}{250} = \frac{y}{100} \qquad y = \frac{25 \times 100}{250} = 10 \qquad \text{This is a 10\% wt/vol solution.}$$

7. *NOTE: We are mixing a gas with water. We assume that there will be no change in volume with the production of the mixture; therefore, we assume that the final volume of this mixture is 750 mL. HINT: 750 mL is 7.5 one hundreds AND percent by weight to volume is calculated by grams per 100 mL. If we divide the 35 g HCl by 7.5, we have 4.67 g per 100 mL solution. That is expressed as 4.67% (wt/vol).*

8. 25% assuming no volume change during production of the solution.

9. 25% sucrose solution requires 50 g of sucrose dissolved to 200 mL final volume.

10. 150 g NaCl and sufficient water to dilute to 1 liter volume.

11. Ratio solutions are produced from a specific number of grams of solute (the left of the ratio) to the final volume of solution. This solution requires 1 gram of $C_{12}H_{22}O_{11}$, sucrose, per 1,000 mL (1 L) of solution.

12. 1 L of this solution requires 0.5 g NaCl. 12 L of the solution is made up using 6 g NaCl.

13. 4 ppm Hg means that there is 4 mg Hg per liter of water taken from the lake. If the fish that comes from the lake is over 1.5 ppm Hg, the fish should not be eaten.

14. 28 parts per thousand (28 $°/_{oo}$)

15. Parts per billion (ppb) is the number of milligrams in 1,000 liters of solution.

16. 9 parts per thousand (9 $°/_{oo}$)

17. 12 grams NaCl are in each liter of the solution.

18. A 0.750 m solution requires 0.750 moles of solute per kg of solvent. The solute in this solution is NaF (MW = 41.988173 g). Therefore, 0.750 moles × 41.9881743 g/mol tells us the mass of the NaF necessary to produce this solution. 31.5 g NaF are dissolved in 1 kg (1 L) of water. The final volume is 1 L since there is no volume change assumed.

19. 0.5 m $C_6H_{12}O_6$

20. 0.17 m Na_2SO_4

21. 5.2 m CH_3OH

22. 1.75 m octane solution in acetone.

23.
$$M = \frac{mol\ NaCl}{L\ solution} = \frac{25\ \cancel{g\ NaCl} \big/ 58\ \cancel{g\ NaCl}/mol}{10\ L\ solution} = 0.043\ M\ NaCl$$

The instructions are to mix the 25 g NaCl with water and dilute to 10 liters final volume while constantly mixing.

24. 0.044 M $C_6H_{12}O_6$

25.
$$M = \frac{mol\ solute}{L\ solution} = \frac{g\ solute \big/ MW}{L\ solution}$$

$$\frac{0.020\ mol}{L} = \frac{y \big/ 78.04554\ g\ Na_2S/mol}{5\ L}$$

$$0.020\ mol \times 5\ \cancel{L} = \frac{y \times \cancel{L}}{78.04554\ g\ Na_2S/mol}$$

$$y = 0.020\ \cancel{mol} \times 5 \times 78.0445\ \cancel{g/mol} = 7.8\ g\ Na_2S$$

Dissolve 6.2 g Na_2S in sufficient water to produce 5 L of solution. Stir during mixing.

26. Weigh 117,695 g H_2SO_4 and dilute to 100 L with distilled or deionized water.

27.
$$M = \frac{mol\ solute}{L\ solution} = \frac{g\ solute\ /\ MW}{L\ solution}$$

$$\frac{0.50\ mol}{1\ L} = \frac{25\ g \big/ 148.31488\ g\ /\ mol}{y}$$

$$0.50\ mol \times y = 1\ L \times \frac{25\ g}{148.31488\ g\ /\ mol}$$

$$y = \frac{1\ L \times \dfrac{25\ \cancel{g}}{148.31488\ \cancel{g}\ /\ \cancel{mol}}}{0.50\ \cancel{mol}} = 0.34\ L$$

28. 198 mL solution mixed (0.198275 L)

29. 26.3 g HCl dissolved in water and diluted to 7.2 liters

30. 7.03 g LiI

31. 2.8 L water, minimum, assuming no volume change during mixing

32. 26.7 mL water

33. 6 N HCl

34. $6 N H_2SO_4$
35. $6 N NH_4OH$. Any volume of the solution is the same concentration; concentration is independent of the volume.
36. $12 N$ NaOH
37. $0.0030 N Ca(OH)_2$
38. The realtionship between normality and volume of the reactants is by $N_1V_1 = N_2V_2$. We substitute into the equation and solve for the answer.

$$V_1 \times 0.10 \, N = 50.0 \, \text{mL} \times 0.150 \, N \qquad V_1 = \frac{50.0 \ \text{mL} \times 0.150 \ N}{0.10 \ N} \qquad V_1 = 75 \, \text{mL}$$

39. 10.27 mL HNO_3
40. Salt (marine) water is hypotonic to the human body. Water contained in the body would tend to move into the intestinal tract. The overall effect is dehydration; the result can be death.
41. The distilled water is hypotonic to muscle cells. Water tends to diffuse from hypotonic to hypertonic. The water will tend to move into the muscle cells; due to the increase in pressure from the water, the cells will expand and may be damaged.
42. In addition to temperature control, some thought must be given to keeping the blood cells in a solution that will maintain the proper tonicity. It turns out that a 0.9% saline (NaCl) solution, **PHYSIOLOGICAL SALINE**, is isotonic with most human tissues. An experiment in which blood cells are placed on a slide under a microscope and exposed to physiological saline would quickly determine if there is any increase or decrease in the volume of the cells.
43. Osmosis is the process by which substances diffuse through a semipermeable membrane. Osmosis is a *passive* process that requires no energy input. The membrane chosen for reverse osmosis is only permeable to water. The problem is to move the water across the membrane and concentrate the water, which is not the normal direction of movement. Reverse osmosis uses an energy source to "pump" the water across the membrane. Hence, the water is concentrated, leaving the salts behind.

10—AQUEOUS EQUILIBRIA

1. Determine the nature of the ionization with a chemical equation, $KCl \rightarrow K^{+1} + Cl^{-1}$. Therefore, $KCl:Cl^{-1}$ is 1:1. We determine the number of moles of KCl we have, which is the same as the number of moles of chloride ions released.

$$M = \frac{\text{mol KCl}}{\text{L solution}} = \frac{\text{g KCl / MW}}{\text{L solution}} = \frac{16.3 \ g / 74.551 \ g / \text{mol}}{1 \ \text{L solution}} = 0.219 \ \text{mol Cl}^{-1} / \text{L}$$

2. 0.63 M hydroxide ions
3. 0.192 moles HCl
4. 1 mole KOH
5. 1.2 M H^{+1}—notice that there are two hydrogen ions for each molecule of H_2SO_4 that ionizes.
6. 32 M OH^{-1}
7. We start this problem by writing the chemical reaction, an equilibrium reaction. We record the information we were given in the problem in the prob. info. row. We know that the reaction is going to progress to the right because we were given zero concentrations for the right side of the reaction. We record these changes in the reaction

row. Adding the prob. info. and the reaction rows provides us with the equilibrium concentrations.

$$HOCN \rightleftharpoons H^{+1} + OCN^{-1}$$

	HOCN	H⁺¹	OCN⁻¹
Prob. Info.	0.10	0	0
Reaction	$-y$	$+y$	$+y$
Equilibrium	$0.10-y$	y	y

We need to determine whether or not y is significant by the 5% rule. The square root of K_a is 0.0187. 5% of 0.10 is 0.005. We cannot eliminate y from the term $0.10 - y$. We write the K_a formula and substitute the problem values into the formula.

$$K_a = \frac{[H^{+1}][OCN^{-1}]}{[HOCN]} \qquad 3.5 \times 10^{-4} = \frac{(y)(y)}{0.10 - y} \qquad \begin{array}{l} y^2 + 3.5 \times 10^{-4}y - 3.5 \times 10^{-5} = 0 \\ y = 5.7 \times 10^{-3}, -6.1 \times 10^{-3} \end{array}$$

The negative value of y won't work. We substitute 5.7×10^{-3} for y into the equilibrium line above and determine the concentration of all species.
$[HOCN] = 0.094$ M; $[H^{+1}] = 0.0057$ M; $[OCN^{-1}] = 0.0057$ M

8. The 5% rule does not apply. 2.3×10^{-3} M F⁻¹.

9. *NOTE: The **25 mL** is information we really do not need. The concentration of a homogeneous solution is the same regardless of the volume of that solution that is sampled.* We write the equation for the disassociation of formic acid and place the information from the problem.

$$HCOOH \rightleftharpoons H^{+1} + HCOO^{-1}.$$

	HCOOH	H⁺¹	HCOO⁻¹
Prob. Info.	0.10	0	0
Reaction	$-y$	$+y$	$+y$
Equilibrium	$0.10 - y$	y	y

$$K_i = \frac{[H^{+1}][HCOO^{-1}]}{[HCOOH]} \qquad 1.8 \times 10^{-4} = \frac{(y)(y)}{0.01 - y} \qquad \begin{array}{l} \text{By the quadratic equation,} \\ [H^{+1}] = 5.8 \times 10^{-3} \end{array}$$

10. The 5% rule applies. 6.0×10^{-7} M PO_4^{-3}

11. The ionization of barium sulfate, $BaSO_4 \rightleftharpoons Ba^{+2} + SO_4^{-2}$, tells us that the concentration of barium ions is the same as sulfate ions. The calculation of the K_{sp} is the product of the ions released (the concentration squared in this problem).
$K_{sp} = [Ba^{+2}][SO_4^{-2}] = (3.317 \times 10^{-3})^2 = 1.1 \times 10^{-5}$

12. 5.0×10^{-15}

13. We write the equation for the ionization, record the K_{sp} formula, substitute in, and solve.
$Bi(OH) \rightleftharpoons Bi^{+3} + 3OH^{-1}$
$K_{sp} = [Bi^{+3}][3OH^{-1}]^3 \qquad K_{sp} = (5.9 \times 10^{-11})(3 \times 5.9 \times 10^{-11})^3 \qquad K_{sp} = 3.3 \times 10^{-40}$

14. $K_{sp} = 1.7 \times 10^{-88}$

15. We write the equation for the ionization of silver chloride and the K_{sp} for that reaction at saturation.
$AgCl_{(s)} \rightleftharpoons Ag^{+1}{}_{(aq)} + Cl^{-1}{}_{(aq)} \qquad K_{sp} = [Ag^{+1}][Cl^{-1}]$
We notice that $[Ag^{+1}] = [Cl^{-1}] = y = AgCl_{(aq)}$, also.
We substitute into the K_{sp} formula and solve for y, which is the solubility of AgCl.
$2.8 \times 10^{-10} = (y)(y) = y^2 \qquad y = 1.7 \times 10^{-5}$ M AgCl in solution at equilibrium.

16. 3.2×10^{-15} molar PbS dissolved at saturation.

17. 1.3×10^{-4} molar Pb^{+2}

18. When we work with solubility product, we assume that whatever amount of the solute dissolves ionizes completely. We write the ionization reaction for magnesium hydroxide in water.

$Mg(OH)_2 \rightleftharpoons Mg^{+2} + 2OH^{-1}$

$K_{sp} = [Mg^{+2}][2OH^{-1}]^2 \quad 1.8 \times 10^{-11} = (y)(2y)^2 = 4y^3 \quad y^3 = 4.5 \times 10^{-12}$

We resolve the cube root by taking the log of 4.5×10^{-12}, dividing by 3, and taking the antilog.

$y = 1.7 \times 10^{-4}$. The solubility of magnesium hydroxide is 1.7×10^{-4} at saturation.

19. 1.4×10^{-5} M

20. We write the equation for the disassociation of the salt. This is a cube root–type problem.

$$Ni(CN)_2 \rightleftharpoons Ni^{+2} + 2CN^{-1}$$
$$ y \qquad 2y$$

We write the K_{sp} formula and solve.

$K_{sp} = [Ni^{+2}][2CN^{-1}]^2 \qquad 3.0 \times 10^{-23} = (y)(2y)^2 = 4y^3 \qquad y^3 = 7.5 \times 10^{-24}$

We resolve the cube root by taking the log of 7.5×10^{-24}, dividing by 3, and taking the antilog. $y = 1.96 \times 10^{-8}$. The solubility of the salt is 1.96×10^{-8} M.

21. $[I^{-1}] = 8.2 \times 10^{-8}$

22. $Sr_3(PO_4)_2 \rightleftharpoons Sr^{+2} + 2PO_4^{-3}$. This is a fifth root situation; we will apply the log method of calculation.

$K_{sp} = [3Sr^{+2}]^3[2PO_4^{-3}]^2 \qquad 1 \times 10^{-31} = (3y)^3(2y)^2 = 108y^5 \qquad y = 2.473519 \times 10^{-7}$

The value of y is the solubility of the salt, 2.5×10^{-7} M.

23. We write the equation for the ionization reaction and place the information from the problem

$$CdCO_3 \rightleftharpoons Cd^{+2} + CO_3^{-2}.$$

Original Solution	0	5.0×10^{-4}
During the Reaction	$+y$	$+y$
Equilibrium	y	5.0×10^{-14}

Application of the 5% rule tells us y is insignificant relative to 5.0×10^{-4}.

We write the K_{sp} formula, substute in, and solve.

$2.5 \times 10^{-14} = (y)(5.0 \times 10^{-4}) \qquad y = 5.0 \times 10^{-11}$

Since one molecule of cadmium carbonate ionizes to release on ion of cadmium (y), the solubility of $CdCO_3$ is 5.0×10^{-11} M.

24. 8.7×10^{-28} M CuS

11—WATER, ACIDS, BASES, AND BUFFERS

1. $pH = -\log[H^{+1}] \quad pH = -\log(0.001) \quad pH = -(-3) = 3$

2. $pH = 1.47$

3. $pH = -\log[H^{+1}] \quad 4.00 = -\log[H^{+1}] \quad \log[H^{+1}] = -4$. Take the antilog of both sides. $[H^{+1}] = 1 \times 10^{-4}$

4. 5.37×10^{-3} M H^{+1}

5. $pOH = pK_w - pH \quad pOH = 14 - 3.2 = 10.8 \quad -\log[OH^{-1}] = 10.8$
   $\log[OH^{-1}] = -10.8$. Take the antilog of both sides. $[OH^{-1}] = 1.6 \times 10^{-11}$

6. 1.78×10^{-9}

7. 7.94×10^{-7}

8. *NOTE: The hydronium ion is a hydrated hydrogen ion, H_3O^{+1}.* 0.178 M

9. $[H_3O^{+1}] = 1 \times 10^{-11} \qquad [OH^{-1}] = 1 \times 10^{-3}$

10. $[H^{+1}] = 7.41 \times 10^{-9} \qquad [OH^{-1}] = 1.35 \times 10^{-6}$

11. $[H^{+1}] = 7.94 \times 10^{-7} \qquad [OH^{-1}] = 1.26 \times 10^{-8}$

12. We write the equation for the equilibrium and substitute in the values.

$$HC_2H_3O_2 \rightleftharpoons H^{+1} + C_2H_3O_2^{-1}$$

Original	0.55	0	0
Reaction	$-y$	$+y$	$+y$
Equilibrium	$0.55 - y$	y	y

We apply the 5% rule and find that we do not have to subtract y from 0.55!

Equilibrium	$0.55 - y$	y	y

$$K_i = \frac{[H^{+1}][HCOO^{-1}]}{[HCOOH]} \quad 1.81 \times 10^{-5} \neq \frac{(y)(y)}{0.55} \quad 9.955 \times 10^{-6} = y^2 \quad y = 3.2 \times 10^{-3}$$

The hydrogen ion concentration is y in the equilibrium line. $[H^{+1}] = 3.2 \times 10^{-3}$

13. $pH = -\log[H^{+1}] \quad pH = -\log 3.2 \times 10^{-3} \quad pH = 2.5$

14. $[H^{+1}] = 0.018$

15. $pH = 1.74$

16. $NH_{3\,(g)} + H_2O_{\,(l)} \rightleftharpoons NH_4^{+1}{}_{(aq)} + OH^{-1}{}_{(aq)}$
    $\quad\;\; 3 - y \quad\;\; - \quad\;\;\;\; y \quad\;\;\;\; y$

    The y is not significant by the 5% rule.

    NOTE: Water is a liquid and is not included in equilibrium calculations.

$$K_b = \frac{[NH_4^{+1}][OH^{-1}]}{[NH_3]} \quad 1.8 \times 10^{-5} = \frac{y^2}{3} \quad y^2 = 5.4 \times 10^{-5} \quad y = 7.3 \times 10^{-3} \text{ M OH}^{-1}$$

17. The pOH is 2.1; therefore, the pH is 11.9.

18. The calculation for a 1:1 ratio buffer system, acid:salt, is $pH = pK_a$.
    $pH = pK_a \qquad pH = -\log 3.5 \times 10^{-4} = 3.5$

19. $pH = 8.6$

20. $HC_2H_3O_2 \rightleftharpoons H^{+1} + C_2H_3O_2^{-1}$

$$pH = pK_a + \log\frac{[\text{conj. base}]}{[\text{acid}]} = pK_a + \log\frac{[C_2H_3O_2^{-1}]}{[HC_2H_3O_2]}$$

$$4.0 = -\log 1.81 \times 10^{-5} + \log\frac{[C_2H_3O_2^{-1}]}{[HC_2H_3O_2]} = 4.7432 + \log\frac{[C_2H_3O_2^{-1}]}{[HC_2H_3O_2]}$$

$$\log\frac{[C_2H_3O_2^{-1}]}{[HC_2H_3O_2]} = -0.7423 \quad \text{Take the antilog of both sides.}$$

$$\frac{[C_2H_3O_2^{-1}]}{[HC_2H_3O_2]} = \frac{0.18}{1} \qquad \text{The salt:acid ratio is 0.18:1 in molarity.}$$

21. The ratio of salt to acid is 2.75 to 1.

22. 1.38 moles $KC_2H_3O_2$

23.
$$HC_2H_3O_2 \rightleftharpoons H^{+1} + C_2H_3O_2^{-1}$$

From Buffer	0.5	K_a	0.5
From NaOH	-0.045		$+0.045$ (Shift right due to water production.)
Equilibrium	0.455	K_a	0.545

$$[H^{+1}] = K_a\frac{[HC_2H_3O_2]}{[C_2H_3O_2^{-1}]} \qquad [H^{+1}] = 1.81 \times 10^{-5} \times \frac{0.455}{0.545} = 1.51 \times 10^{-5}$$

24. $[H^{+1}] = 3.3 \times 10^{-10}$

25. Buffer systems do not function beyond ± 1.00 pH units from the pK (pK_b if a base and

pK$_a$ if an acid). Buffer systems become exhausted if applied beyond the limits.

Chapter 11
Chapter 12

12—COLLIGATIVE PROPERTIES

NOTE: Since we are not given the density of water for any of the problems in this chapter, we assume the density of one gram per millilter.

1. (a)

$$m = \frac{{}^g\!/_{MW}}{kg} = \frac{{}^{85\,g}\!/_{180.1572\,g/mol}}{1\,kg} = 0.47\ m$$

(b) $\Delta T_f = i \times k_f \times m$ *NOTE: i is 1 for organic solutes, they don't ionize.*
$\Delta T_f = 1 \times 1.86\ {}^{\circ}C/_m \times 0.47\ m$
$\Delta T_f = 0.88\,^{\circ}C$
and
$FP = FP^{\circ} - \Delta T_f$
$FP = 0\,^{\circ}C - 0.88\,^{\circ}C = -0.88\,^{\circ}C$

2. (a) 0.175 m (b) −0.33°C
3. (a) 0.706 m (b) 13.8°C
4. (a) 0.139 m (b) 100.07°C
5. (a) 0.333 m (b) 78.91°C
6. (a) 0.333 m (b) −1.24°C (c) 100.17°C
7. $\Delta T_f = i \times k_f \times m$ *NOTE: $i = 5$. $Al_2(SO_4)_3 \rightarrow 2Al^{+3} + 3SO_4^{-2}$.*

$$\Delta T_f = 5 \times 1.86\,^{\circ}C/_m \times \frac{{}^{45\,g}\!/_{342.15388\,g/mol}}{3\,kg} = 0.4077\,^{\circ}C$$

$FP = FP^{\circ} - \Delta T_f$
$FP = 0\,^{\circ}C - 0.41\,^{\circ}C = -0.41\,^{\circ}C$

8. 212.19°C

9. $$X_{glucose} = \frac{mol\ glucose}{mol\ glucose + mol\ water} = \frac{{}^{50}\!/_{180.1572}}{{}^{50}\!/_{180.1572} + {}^{250}\!/_{18.0152}} = 0.0196$$

10. 0.0104

11. In a sense, we are playing a mathematical game with these two problems. The concentration of the glucose is greater than that of the sucrose because the molecular weight of glucose is lower than that of sucrose. Even though the denominators of the fractions are similar, the numerators are significantly different.

12. $$P_{solv.} = X_{solv.} \times P_{solv.} = \frac{{}^{300}\!/_{18.0152}}{{}^{32}\!/_{180.1572} + {}^{300}\!/_{18.0152}} \times 31.824 = 31.5\ mm\ Hg$$

13. The vapor pressure is 758 mm Hg; standard vapor pressure is 760 mm Hg. The reduction in the vapor pressure increases the boiling point above the standard 100°C.

14. $$\pi = MRT = \frac{mol\ solute \times R \times T}{L\ solution}$$

$$\frac{3\ mol \times 0.0821\ {}^{L\,atm}\!/_{mol\,K} \times 298K}{L} = 73.4\ atm$$

224

*Solutions
and
Answers*

Chapter 12
Chapter 13

15. 2.83 atm

16. $P_{solvent} = \dfrac{moles\ water}{moles\ water\ +\ moles\ solute} \times P^{o}$

$22.98 = \dfrac{85/18.0152}{85/18.0152\ +\ 7.5/MW} \times 23.76$

MW= 46.8 g/mol

17. $\Delta T_{b} = i \times k_{b} \times m$ $\qquad \Delta T_{b} = \dfrac{i \times k_{b} \times \dfrac{g\ solute}{MW}}{kg\ solvent}$

$MW = \dfrac{i \times k_{b} \times g\ solute}{\Delta T_{b} \times kg\ solvent}$ $\qquad MW = \dfrac{1 \times 0.512 \times 150}{2.3 \times 0.650} = 51.4$ g/mol

18. $MW = \dfrac{1 \times 1.86 \times 56}{3.45 \times 0.100} = 301.9$ g/mol

13—NUCLEAR CHEMISTRY

1. $^{210}_{89}Ac \rightarrow ^{206}_{87}Fr + ^{4}_{2}He$ $\qquad$ 2. $^{212}_{86}Rn \rightarrow ^{208}_{84}Po + ^{4}_{2}He$

3. $^{116}_{47}Ag \rightarrow ^{112}_{45}Rh + ^{4}_{2}He$ $\qquad$ 4. $^{231}_{90}Th \rightarrow ^{231}_{91}Pa + \beta^{-}$

5. $^{206}_{80}Hg \rightarrow ^{206}_{81}Tl + \beta^{-}$ $\qquad$ 6. $^{214}_{82}Pb \rightarrow ^{214}_{83}Bi + \beta^{-}$

7. $^{202}_{82}Pb \rightarrow ^{202}_{81}Tl + \beta^{+}$ $\qquad$ 8. $^{131}_{53}I \rightarrow ^{131}_{52}Te + \beta^{+}$

9. $^{131}_{54}Xe \rightarrow ^{131}_{53}I + \beta^{+}$

10. $^{131}_{54}Xe \rightarrow ^{127}_{52}Te + ^{4}_{2}He \rightarrow ^{127}_{53}I + \beta^{-}$

11. $^{209}_{84}Po \rightarrow ^{209}_{85}Ac + \beta^{-} \rightarrow ^{209}_{86}Rn + \beta^{-}$

12. $^{251}_{98}Cf \rightarrow ^{247}_{96}Cm + ^{4}_{2}He \rightarrow ^{243}_{94}Pu + ^{4}_{2}He \rightarrow ^{243}_{95}Am + \beta^{-}$

13. $^{56}_{26}Fe \rightarrow ^{52}_{24}Cr + ^{4}_{2}He \rightarrow ^{52}_{25}Mn + \beta^{-} \rightarrow ^{52}_{26}Fe + \beta^{-} \rightarrow ^{48}_{24}Cr + ^{4}_{2}He$

14. $^{110}_{47}Ag \rightarrow ^{110}_{48}Cd + \beta^{-}$ $\qquad$ 15. $^{37}_{16}S \rightarrow ^{37}_{17}Cl + \beta^{-}$

16. $^{235}_{93}Np \rightarrow ^{235}_{92}U + \beta^{+}$ $\qquad$ 17. 424,000 years

18. 5 half-lives—28.75 years $\qquad$ 19. Slightly over six half-lives

20. I-131 is the better choice because it will remain in the thyroid long enough to damage the cancer cells and it is both a beta and gamma emitter. I-123 has a much shorter half-life and, therefore, is not present in the thyroid gland for nearly as long. Furthermore, I-123 is only a gamma emitter.

21. A full-body scan would show the uptake of Tc-99 by cancerous tissue. There would be what is called a *hot spot* where the radioisotope accumulates—the cancerous tissue in the liver. Another hot spot shows at the injection site, which is normally at the elbow. The reason for the full-body scan is that the scan indicates if the cancer has metastasized.

22. Kr-81, as a radioisotope, can damage tissue. However, the half-life of Kr-81 is long

enough that the chances of any cell injury are small; the number of disintegrations per minute is low. Kr-81 is introduced into the lungs and totally exhaled in a short period of time; this reduces the exposure to the radioisotope. Furthermore, krypton is an inert gas. Inert gases do not react chemically and are not retained by the body. All of these factors reduce the statistical probability of damage to tissue.

23. We set up this problem so that we express the amount of exposure in one day. Then calculate to see how many days of that exposure it takes to reach 50 R.

$$\frac{3\ R}{2\ days}\ y = 50\ R$$

$$3\ R\ y = 50 \times 2\ R\ days$$

$$y = \frac{100\ \cancel{R}\ days}{3\ \cancel{R}} = 33.3\ days$$

24. This person is receiving 50 R in 33.3 days. Assuming an exposure every day for a year, the person receives 365/33.3 exposures.

$$\frac{50\ R}{exposure} \times \frac{365\ days/year}{33.3\ days/exposure} = 548\ \frac{R}{yr}$$

We convert R to rads by the multiplication of R by 0.96. This is an exposure of 526 rad per year. This individual is in trouble. An exposure of 200 rads brings about radiation sickness; 450 rads is enough for one-half of the humans exposed to die.

25. The conversion from roentgens to rads is by 0.96 $^{rad}/_R$. Since the federal guidelines are in rads, we convert.

$$2.3\ \cancel{R} \times 0.96\ ^{rad}/_{\cancel{R}} = 2.208\ rad\ exposure\ for\ one\ week$$

We extend the exposure of one week to one year.

$$2.208\ ^{rad}/_{wk} \times 50\ ^{wk}/_{yr} = 110.4\ ^{rad}/_{year}$$

An exposure of 50 rad in one year results in no radiation sickness and usually causes no damage to gamete-producing tissue that would lead to mutations in the gametes. An exposure of 200 rad generally results in radiation sickness and possible long-term effects. This worker is being exposed to much more than is acceptable, and the constant exposure over time increases the probability of damage to healthy tissue.

14—HYDROCARBON FAMILIES

1. We notice that this compound is symmetrical and that we can name either terminal carbon as the number 1 carbon. The name of this compound is 1,2,3-trichloro 2-iodo propane.

2. 3-chloro 1,1-difluoro 4-iodo pentane

3. The longest chain is three carbons. Don't be fooled by the angular sketch—remember that the true appearence of the molecule is on the basis of 109°30' angles from the carbon(s). This compound is propane.

4. 1,4-dichloro 2-methyl butane

5.
```
     H  H  H  F  H
     |  |  |  |  |
Br – C– C– C– C– C– H
     |  |  |  |  |
     H  F  H  H  H
```

6.
```
     Cl H
     |  |
Cl – C– C– F
     |  |
     H  H
```

226

*Solutions
and
Answers*

Chapter 14
Chapter 15

7.
```
    H  F  H  F  H
    |  |  |  |  |
H — C — C — C — C — C — H
    |  |  |  |  |
    I  I  H  F  H
```

8.
```
   Cl  F  H  H  H Cl
    |  |  |  |  |  |
H — C — C — C — C — C — C — H
    |  |  |  |  |  |
   Cl  F  C  H  C Cl
          |     |
          H     H
```

9.
```
    H  H Cl  H  H  H  H  H
    |  |  |  |  |  |  |  |
H — C — C — C — C — C — C — C — C — H
    |  |  |  |  |  |  |  |
    H  H Cl H C H H Cl  H  H
            |
            H C H
              |
              H
```

10.
```
    H  H  H  I  H  H  H Cl Cl
    |  |  |  |  |  |  |  |  |
H — C — C — C — C — C — C — C — C — C — Cl
    |  |  |  |  |  |  |  |  |
    H  H  I  H  H Cl  H  I  H
```

11.
```
    H  Br  I  CH₃ H  Br
    |  |   |  |   |  |
H — C = C — C — C — C — C — H
    |  |   |  |   |  |
    H  H   H  H  CH₃ H
```

12.
```
    H  H  H  H Cl  H  H  H
    |  |  |  |  |  |  |  |
H — C = C — C = C — C — C — C — C — H
                |  |  |  |
                H  Br H  H
```

13. There are three isomers of hexane substituted by one fluorine.

```
    F  H  H  H  H  H        H  F  H  H  H  H        H  H  F  H  H  H
    |  |  |  |  |  |        |  |  |  |  |  |        |  |  |  |  |  |
H — C — C — C — C — C — C — H   H — C — C — C — C — C — C — H   H — C — C — C — C — C — C — H
    |  |  |  |  |  |        |  |  |  |  |  |        |  |  |  |  |  |
    H  H  H  H  H  H        H  H  H  H  H  H        H  H  H  H  H  H
```

14. There is only one structure of ethyl hexane, 2-ethyl hexane. The reason is that the ethyl group cannot be substituted on the number 2 carbon because we end up with a seven-carbon compound.

15. We cannot place a methyl group on a terminal carbon because we end up with a nine-carbon skeleton if we do. There are three isomers of methyl octane.

16. There are three isomers of ethane substituted with one fluorine and one chlorine.

17. Don't forget *cis* and *trans* structures. There are five isomers of difluoro propene.

18. There are three isomers of *trans*-diflouro hexene.

```
    F  H  H  H  H  H        H  F  H  H  H  H        H  H  F  H  H  H
    |  |  |  |  |  |        |  |  |  |  |  |        |  |  |  |  |  |
H — C = C — C — C — C — C — H   H — C — C = C — C — C — C — H   H — C — C — C = C — C — C — H
    |  |  |  |  |  |        |  |  |  |  |  |        |  |  |  |  |  |
    H  F  H  H  H  H        H  H  F  H  H  H        H  H  H  F  H  H
```

19. 5-chloro 5,5-difluro *trans*-1,2 *cis*-3,4-tetraiodo 1,3-diene pentane

20. 1-chloro 1,4,8,8-tetrafluoro *trans*-5,6-diiodo 4-methyl 5-ene 2-yne octane

15—ALCOHOLS, ACIDS, ESTERS, AND ETHERS

1. The alcohol function group is similar to the hydroxide group.

2. The carbons in organic compounds are bound by means of covalent bonds. The alcohol group is covalently bound to the carbon, also. This sharing of electrons does not allow the oxygen in the functional group to attract electrons. On the other hand, sodium hydroxide is not covalently bound. The oxygen in the hydroxide can accumulate electrons from the sodium, leading to an ionic bond.

3. A primary alcohol is a compound with the alcohol group on a terminal carbon of the longest chain. A secondary alcohol is a compound with the alcohol group on an interior carbon of the longest chain.

4.
```
    H  H  H  H
    |  |  |  |
H — C — C — C — C — H
    |  |  |  |
    H  OH H  H
```
This compound is a secondary alcohol.

5.
```
    H  H  H  H  H
    |  |  |  |  |
H — C — C — C — C — C — H
    |  |  |  |  |
    H  H  OH H  H
```
This compound is a secondary alcohol.

6. An alcohol is a compound with the general formula of **R−O−H**, whereas an organic acid is

$$\underset{\text{R}-\overset{\overset{\displaystyle O}{\|}}{C}-O-H}{}$$

7. The carboxyl carbon only has one location (R, see answer 5) to link. If the carboxyl carbon is linked to an interior carbon, there are too many chemical bonds.

8. Alcohols and organic acids are expected to react by means of a dehydration reaction. The hydrogen comes from the acid and the alcohol group leaves the compound to hook up with the hydrogen. The final product of the portion of the reaction is water. The remaining functional groups link to produce an ester, an organic salt.

9.

$$\underset{\substack{| \quad | \\ \text{H} \ \text{H}}}{\overset{\substack{\text{H} \ \text{H} \ \text{O} \\ | \quad | \quad \|}}{\text{H}-\text{C}-\text{C}-\text{C}-\text{O}-\text{H}}} + \underset{\substack{| \quad | \quad | \\ \text{H} \ \text{H} \ \text{H}}}{\overset{\substack{\text{H} \ \text{H} \ \text{H} \\ | \quad | \quad |}}{\text{H}-\text{O}-\text{C}-\text{C}-\text{C}-\text{H}}} \rightarrow \underset{\substack{| \quad | \\ \text{H} \ \text{H}}}{\overset{\substack{\text{H} \ \text{H} \ \text{O} \\ | \quad | \quad \|}}{\text{H}-\text{C}-\text{C}-\text{C}-\text{O}}}-\underset{\substack{| \quad | \quad | \\ \text{H} \ \text{H} \ \text{H}}}{\overset{\substack{\text{H} \ \text{H} \ \text{H} \\ | \quad | \quad |}}{\text{C}-\text{C}-\text{C}-\text{H}}} + \text{H}-\text{O}-\text{H}$$

10. The compound produced, other than water, is an ester. The name of the compound is propyl propanate.

11.

$$\underset{\substack{| \quad | \quad | \quad | \\ \text{H} \ \text{H} \ \text{H} \ \text{H}}}{\overset{\substack{\text{H} \ \text{H} \ \text{H} \ \text{H} \\ | \quad | \quad | \quad |}}{\text{H}-\text{C}-\text{C}-\text{C}-\text{C}-\text{O}-\text{H}}} + \underset{\substack{| \\ \text{H}}}{\overset{\substack{\text{O} \ \text{H} \\ \| \quad |}}{\text{H}-\text{O}-\text{C}-\text{C}-\text{H}}} \rightarrow \underset{\substack{| \quad | \quad | \quad | \\ \text{H} \ \text{H} \ \text{H} \ \text{H}}}{\overset{\substack{\text{H} \ \text{H} \ \text{H} \ \text{H} \\ | \quad | \quad | \quad |}}{\text{H}-\text{C}-\text{C}-\text{C}-\text{C}-\text{O}}}-\underset{\substack{| \\ \text{H}}}{\overset{\substack{\text{O} \ \text{H} \\ \| \quad |}}{\text{C}-\text{C}-\text{H}}} + \text{H}-\text{O}-\text{H}$$

12. The ester produced is butyl acetate.

13. This compound is an ester. The name is ethyl acetate. The ester was produced by the dehydration reaction of ethyl alcohol and acetic acid.

14. This compound is butyl propanate and is an ester. The reactants that produced this ester are *n*-butanol and propanoic acid.

15. Lipids are produced by an ester linkage between three fatty acids and one glycerol. The inorganic reaction that is analogous is an acid-base reaction.

16. We can write the formula for nitric acid to indicate how it reacts rather than using the formula, HNO_3.

$$\begin{array}{l}\underset{\substack{| \\ \text{H}}}{\overset{\substack{\text{H} \\ |}}{\text{H}-\text{C}\text{-}\!\!\!\fbox{O}\text{-H} \quad \fbox{H}\text{-O}-\text{NO}_2}} \\ \text{H}-\text{C}\text{-}\!\!\!\fbox{O}\text{-H} \quad \fbox{H}\text{-O}-\text{NO}_2 \\ \underset{\substack{| \\ \text{H}}}{\text{H}-\text{C}\text{-}\!\!\!\fbox{O}\text{-H} \quad \fbox{H}\text{-O}-\text{NO}_2}\end{array} \rightarrow \begin{array}{l}\underset{\substack{| \\ \text{H}}}{\overset{\substack{\text{H} \\ |}}{\text{H}-\text{C}-\text{O}-\text{NO}_2}} \\ \text{H}-\text{C}-\text{O}-\text{NO}_2 \\ \underset{\substack{| \\ \text{H}}}{\text{H}-\text{C}-\text{O}-\text{NO}_2}\end{array} + 3\,\text{H}-\text{O}-\text{H}$$

17. A saturated fat is one that has no double or triple bonds between the carbons of the fatty acid fractions. An unsaturated fat does have one or more double and/or triple bonds.

18. The original oil is an unsaturated lipid to which hydrogen has been added. The double (or triple) bond(s) is opened and hydrogen is placed.

19. There is a negative pole located at the oxygen of an alcohol; however, it is a weak pole compared the negative center of the acid's doubly bound oxygen. Furthermore, the acid structure also has a (weaker) negative center located at the singly bound oxygen.

$$\text{Alcohol:} \ \underset{\substack{| \\ \text{H}}}{\text{R}-\text{O}} \qquad \text{Acid:} \ \text{R}-\overset{\overset{\displaystyle O}{\|}}{\text{C}}-\underset{\substack{| \\ \text{H}}}{\text{O}}$$

20. We can compare the ether linkage, **R−O−R**, with the ester linkage. However, let's use dimethyl ether so we can see how the carbons are situated.

Dimethyl ether, $\underset{\substack{| \quad | \\ \text{H} \ \text{H}}}{\overset{\substack{\text{H} \ \text{H} \\ | \quad |}}{\text{H}-\text{C}-\text{O}-\text{C}-\text{H}}}$, and the ester linkage, $\underset{\substack{| \\ \text{H}}}{\text{R}'-\overset{\overset{\displaystyle O}{\|}}{\text{C}}-\text{O}-\overset{\substack{\text{H} \\ |}}{\text{C}}-\text{R}''}$, differ with the

228

*Solutions
and
Answers*

Chapter 15
Chapter 16

presence of the doubly bound oxygen found in the ester and not found in the ether. This oxygen is a negative center and polar. The ether linkage has no polarity similar to that of the ester linkage.

21. This compound is ethyl propyl ether.

$$
\underset{\textbf{Ethanol}}{H-\overset{\overset{H}{|}}{\underset{\underset{H}{|}}{C}}-\overset{\overset{H}{|}}{\underset{\underset{H}{|}}{C}}-O-H}
\quad
\underset{\textbf{Propanol}}{H-O-\overset{\overset{H}{|}}{\underset{\underset{H}{|}}{C}}-\overset{\overset{H}{|}}{\underset{\underset{H}{|}}{C}}-\overset{\overset{H}{|}}{\underset{\underset{H}{|}}{C}}-H}
\overset{\overset{Conc.}{H_2SO_4}}{\longrightarrow}
\underset{\textbf{Ethyl Propyl Ether}}{H-\overset{\overset{H}{|}}{\underset{\underset{H}{|}}{C}}-\overset{\overset{H}{|}}{\underset{\underset{H}{|}}{C}}-O-\overset{\overset{H}{|}}{\underset{\underset{H}{|}}{C}}-\overset{\overset{H}{|}}{\underset{\underset{H}{|}}{C}}-\overset{\overset{H}{|}}{\underset{\underset{H}{|}}{C}}-H}
+
\underset{\textbf{Water}}{H-O-H}
$$

22. This compound is methyl isopropyl ether and was produced by the dehydration reaction of methanol and isopropanol.

16—OTHER CLASSES OF ORGANIC COMPOUNDS

1. Butaldehyde (butanal)
2. Formaldehyde, methanal

3.
$$
H-\overset{\overset{H}{|}}{\underset{\underset{H}{|}}{C}}-\overset{\overset{H}{|}}{\underset{\underset{H}{|}}{C}}-\overset{\overset{H}{|}}{\underset{\underset{H}{|}}{C}}-\overset{\overset{H}{|}}{\underset{\underset{H}{|}}{C}}-\overset{\overset{O}{\|}}{C}-H
$$

4.
$$
H-\overset{\overset{H}{|}}{\underset{\underset{H}{|}}{C}}-\overset{\overset{O}{\|}}{C}-H
$$

The carbonyl carbon is the right-hand carbon because it is doubly bound to an oxygen.

5. Propyl ketone (4-heptanone)
6. Methyl hexyl ketone; 2-octanone

7.
$$
H-\overset{\overset{H}{|}}{\underset{\underset{H}{|}}{C}}-\overset{\overset{O}{\|}}{C}-\overset{\overset{H}{|}}{\underset{\underset{H}{|}}{C}}-\overset{\overset{H}{|}}{\underset{\underset{H}{|}}{C}}-\overset{\overset{H}{|}}{\underset{\underset{H}{|}}{C}}-H
$$
The IUPAC name is 2-pentanone.

8. The only polar point on this compound is the doubly bound oxygen. It is negative.
9. An amine is a compound with an –NH_2 group. An amide is a compound with a carbonyl carbon and an –NH_2 group. An amino acid is an example of an amide in which there is a carboxyl carbon and an amine group on a different carbon.
10. Polar points of opposite charge are attracted to each other and form hydrogen bonds.
11. $R-\overset{}{\underset{\underset{H}{|}}{S}}$ The sulfur tends to be negative and the hydrogen tends to be positive.

12. The disulfide linkage is not polar because the two sulfurs are in the middle of the compound. Essentially, the sulfurs are buried in the compound and not exposed to the environment of the compound.

13.

14. *p*-bromochlorobenzene
15. *m*-bromochlorobenzene
16. bromo *o*-dichlorobenzene
17. 1,2,2-tricholorbenzene
18. 2,3,5-trifluorophenol
19. *p*-bromoethylbenzene
20. 2,3,5-trifluorotoluene
21. The two structures are isomers. (The isomers are of benzpyrene; both isomers are carcinogenic. They are byproducts of cigarette smoke, barbeque cooking, and burning wood.)
22. All of the carbohydrates mentioned have the same general formula, $C_6H_{12}O_6$, These compounds are isomers.
23. Lactose is bound by a beta bond. Beta bonds are more difficult to digest than alpha bonds in general.
24. A dehydration reaction can be performed between the alcohol on the left side of glucose, the number 3 carbon, and the alcohol, $-CH_2OH$, on the right side of the fructose, the number 6 carbon. The bond produced is an alpha bond.
25. (a) carbonyl carbon
    (b) alchohol group
    (c) methyl group
    (d) This carbon is a cabonyl carbon (ketone). The oxygen is a negative pole on the compound.
    (e) $-CH_2OH$ was derived from CH_3OH, methanol.
    (f) $-CH_3$ was derived from CH_4, methane.

INDEX/GLOSSARY

*The page numbers are the first time that the term is used/defined
and/or the page with the most comprehensive use or definition.
Those terms that do not have a page number are general usage
terms; the definition is provided for your convenience.*

are removed. *See* Molecular Formula.

Equality	—	In an equation, the left equals the right side.
Equilibrium	95	A chemical or physical reaction in which there is movement in opposite directions at the same rate.
Equilibrium Constant	95	A mathematical expression that is the ratio of the product of the products to the product of the reactants.
Ester	183	An organic salt; the product of an acid-alcohol reaction.
Ether	185	An organic compound with two functional groups bound by an oxygen, R–O–R.
Factor	3	One of two or more numbers that are multiplied.
Free Element	—	An element that is free in nature. The element is not combined with others.
Freezing Point	—	(FP) The temperature at which a substance (pure substance or mixture) solidifies under the condition of one atmosphere.
Freezing Point Lowering	147	(ΔT_f) The decrease in freezing point proportional to the molality of the solute particles in solution.
Gram Atomic Weight	43	The atomic weight expressed in grams
Gray	166	(Gy) 100 rad, the transfer of 240,000 calories of energy to 1 kg of tissue.
Group	19	Columns found on the Periodic Table of the Elements.
Guesstimate	—	The estimation of the size of an answer before calculation.
Half-Life	161	The time period for one-half of a sample's nuclei to decay.
Half Reaction	63	An oxidation or a reduction reaction. Half reactions can be used to balance a redox reaction.
Halogen	57	Group VIIA nonmetals: fluorine, chlorine, bromine, and iodine.
Heavy Water	24	Water with the molecular weight higher than the $18 \, ^g/_{mol}$ predicted by the Periodic Table of the Elements due to deuterium and/or tritium instead of hydrogen.
Henderson-Hasselbalch Equation	142	The mathematical relationship used to calculate weak acid or base:strong salt ratio to achieve a pH other than that of pK_a or pK_b.
Heterocyclic Compound	194	Ring compounds with one or more of the carbons replaced by another element, commonly oxygen or nitrogen.
Heterogeneous	—	Members are different from each other. A chemical reaction with members in different physical states (i.e., a solid that decomposes to a solid and a gas).
Homogeneous	—	All members are the same. A chemical reaction with members all in the same physical state (i.e., a solid that decomposes to release two other solids). Solutions do not separate; solutions are homogeneous mixtures.
Hydrocarbon	—	A compound of only hydrogen and carbon.
Hydrogenated	185	Reintroduction of hydrogens into a chemical bond, either double or triple.